LAB MANUAL TO ACCOMPANY

Agricultural Mechanics

Join us on the web at

www.cengage.com/community/agriculture

LAB MANUAL TO ACCOMPANY
Agricultural Mechanics
Fundamentals and Applications

7TH edition

Ray V. Herren, EdD

CENGAGE Learning®

Australia • Brazil • Japan • Korea • Mexico • Singapore • Spain • United Kingdom • United States

CENGAGE Learning

Lab Manual to Accompany Agricultural Mechanics: Fundamentals and Applications, 7th Edition

Ray V. Herren, EdD

General Manager: Dawn Gerrain

Associate Product Manager: Nicole Sgueglia

Senior Director, Development: Marah Bellegarde

Senior Product Development Manager: Larry Main

Senior Content Developer: Laura J. Stewart

Product Assistant: Scott Royael

Executive Director of Marketing: Carolyn Pestritto

Sr. Product Marketing Manager: Leah Klein

Senior Production Director: Wendy A. Troeger

Production Manager: Mark Bernard

Senior Content Project Manager: Elizabeth C. Hough

Senior Art Director: David Arsenault

Cover image credits: Irrigation Equipment: © Steven Ringler/www.Shutterstock.com

Credits for top row of images: (top left image) Saw: © Cengage Learning 2015; (top middle image) Electrical Outlet: © Brandon Blinkenberg/www.Shutterstock.com; (top right image) Linkage: © iStockphoto/francv

DESIGN IMAGE CREDITS

Section opener image credit: © Steven Ringler/www.Shutterstock.com

Unit Opener: Unit 01— © iStockphoto/Simply Creative Photography, Unit 02: © iStockphoto/asterix0597, Units 03-06: Copyright © 2015 Cengage Learning®., Unit 07: © iStockphoto/flyfloor, Unit 08: © iStockphoto/Zmiy, Units 09-16: Copyright © 2015 Cengage Learning®., Unit 17: © iStockphoto/Prill Mediendesign & Fotografie, Unit 18: © iStockphoto/Ruslan Dashinsky, Unit 19: © iStockphoto/taonga, Unit 20: © iStockphoto/Silvia Jansen, Units 21-22: Copyright © 2015 Cengage Learning®., Unit 23: © iStockphoto/GlenJ, Unit 24: Copyright © 2015 Cengage Learning®. Unit 25: © Larry Jeffus, Unit 26: © iStockphoto/kentarus, Unit 27: Copyright © 2015 Cengage Learning®, Unit 28: © iStockphoto/kai813, Unit 29: © iStockphoto/Janice Richard, Unit 30: Copyright © 2015 Cengage Learning®, Unit 31: © iStockphoto/Banks Photos, Unit 32: © iStockphoto/daseaford, Unit 33: © iStockphoto/Fertnig, Unit 34: © iStockphoto/Gary Alvis, Unit 35: © iStockphoto/Living Images, Unit 36: © iStockphoto/emesilva, Unit 37: © iStockphoto/Branislav, Unit 38: © iStockphoto/cenix, Unit 39: © iStockphoto/Banks Photos, Unit 40: © iStockphoto/Carole Gomez, Unit 41: © iStockphoto/iacona, Unit 42: Copyright © 2015 Cengage Learning®, Unit 43: © iStockphoto/Charles Schug

Library of Congress Control Number: 2013943110
ISBN-13: 978-1-285-05901-3

Cengage Learning
200 First Stamford Place, 4th Floor
Stamford, CT 06902
USA

Cengage Learning is a leading provider of customized learning solutions with office locations around the globe, including Singapore, the United Kingdom, Australia, Mexico, Brazil, and Japan. Locate your local office at: **www.cengage.com/global**

Cengage Learning products are represented in Canada by Nelson Education, Ltd.

To learn more about Cengage Learning, visit **www.cengage.com**
Purchase any of our products at your local college store or at our preferred online store **www.cengagebrain.com**

Notice to the Reader

Printed in the United States of America
1 2 3 4 5 6 7 18 17 16 15 14

Contents

List of Figures

Introduction

This lab manual was designed to accompany the core textbook, *Agricultural Mechanics: Fundamentals and Applications*, 7th Edition. For each unit, the lab manual provides class and shop activities that will help you learn and apply the concepts discussed. Many of the activities relate to procedures and tasks you will learn in the shop, and the class activities are designed to help you recall important knowledge about each topic. Along the way, your instructor will indicate scores or approval for each assigned activity.

SECTION 1

Exploring Careers in Agricultural Mechanics

Unit 1

Mechanics in the World of Agriculture

Unit 2

Career Options in Agricultural Mechanics

UNIT 1

Mechanics in the World of Agriculture

Many jobs and careers are available in the field of agricultural mechanics. This unit is designed to help you explore some of them.

Class Activity 1-1: Studying Agricultural Mechanics Careers

Job Sheet 1-1

Name ______________________________ Date ____________

Studying Agricultural Mechanics Careers

Objective

Upon completing this activity, you will have demonstrated an understanding of agricultural mechanics careers.

Tools and Materials

Job Sheet 1-1

Textbook

Pen or pencil

Procedure

Answer the following questions:

1. Define agriculture. ______________________________

____

2. Which of the following are agricultural products?

	Yes	*No*
Paints	___	___
Plastics	___	___
Tar	___	___
Fertilizer	___	___
Glass	___	___
Leather	___	___
Silicone	___	___
Sod	___	___

3. Why is agriculture called a basic industry?

____

____

____

4. How many people does our textbook suggest the average farmer feeds?

____

5. Why are renewable natural resources considered to be part of the agriculture industry?

____

____

____

6. Define efficiency. ______________________________

____

7. Due to the efficiency of today's agriculture industry, only ___ out of every ___ Americans works in agriculture.
8. What is the role of agricultural mechanics in America's productivity: major or minor? (Circle the best answer.)
9. List five jobs requiring agricultural mechanics skills:
____
____
____
____
____
10. Why was John Deere's improvement of Thomas Jefferson's iron plow so important?
____
____
____
11. Why was Whitney's cotton gin so important in the 1790s?
____
____
____
12. Many other mechanical inventions were important because they improved _______________
____
13. Why are American machines not widely used in other countries?
____
____
____
14. Can you name one use of computers in agriculture?
____
15. How might the use of computers and robotics influence agriculture in the future?
____
____
____

Instructor's Score or Approval _______________

UNIT 2

Career Options in Agricultural Mechanics

Unit 2 continues the study of agricultural occupations.

Class Activity 2-1: Agricultural Occupations

Job Sheet 2-1

Name ______________________________ Date ____________

Agricultural Occupations

Objective

Upon completing this activity, you will have demonstrated an understanding of agricultural occupations.

Tools and Materials

Job Sheet 2-1

Textbook

Pen or pencil

Procedure

Answer the following questions:

1. Off-the-farm agricultural jobs are those requiring ______________ but are not farming or ranching.
2. What attracts many people to careers in agriculture?

   ____

   ____
3. Americans use approximately ______________ percent of their total disposable income for food.
4. How does this percentage compare with other parts of the world?

   ____
5. What are the levels of employment in agriculture?

   ____
6. What percentage of employment in agriculture is in production?

   ____
7. What education is required for the professional employment level?

   ____

   ____
8. List the eight divisions of agribusiness and agricultural production.

   ____

   ____

   ____

   ____

   ____

   ____

   ____

   ____
9. List the six divisions in agricultural science.

   ____

   ____

10. List the seven divisions in renewable natural resources.

11. What are the specific job title categories in agricultural mechanics?

12. List some questions you should ask yourself when looking into career choices.

13. How can you prepare yourself for a career in the field of agriculture?

14. Which student organizations can help you develop useful skills for agricultural or other occupations?

Instructor's Score or Approval ______________________

SECTION 2

Using the Agricultural Mechanics Shop

UNIT 3

Shop Orientation and Procedures

The shop offers you many hands-on opportunities to learn skills useful to you in many of the careers discussed in the first two units. To get the most from your experiences, you need to become familiar with the shop, tools, and safety.

The activities in this unit are designed to help you identify the capabilities of your shop and become acquainted with its tools and work areas. Complete each activity with the goal of gaining the most knowledge you can from your shop experience.

Class Activity 3-1: Identifying the Importance of Agricultural Mechanics Activities

Shop Activity 3-2: Sketching and Labeling the Agricultural Mechanics Shop

Shop Activity 3-3: Completing the Shop Policies and Procedures Form

Class Activity 3-4: Completing the Allergies and Physical Problems Statement

Job Sheet 3-1

Name ______________________________________ Date ______________

Identifying the Importance of Agricultural Mechanics Activities

Objective

Upon completing this activity, you will have demonstrated the ability to identify the importance of agricultural mechanics activities.

Tools and Materials

Job Sheet 3-1

Pen or pencil

Procedure

Using the textbook as a reference, answer the following questions:

1. Complete the following statement: The agricultural mechanics shop is an important learning place because
____
____
____
____
____.

2. List the categories of shop operations given in the textbook.
____
____
____
____

3. Why are open areas important in the shop?
____
____
____
____

4. Why are electrical skills important in agricultural mechanics?
____
____

5. What agricultural power machinery experiences are named in the textbook that students might encounter in the shop?
____
____
____

6. The agricultural mechanics shop should be adequate in size for safe and efficient work by ____________ high school students.

7. Tell why you agree or disagree with the shop policies and procedures in Figure 3-15 in the textbook.

____

____

____

____

____

____

____

____

Instructor's Score or Approval ______________________________

Job Sheet 3-2

Name ______________________________________ Date ______________

Sketching and Labeling the Agricultural Mechanics Shop

Objective

Upon completing this activity, you will have demonstrated the ability to identify and sketch the major work areas, stationary power tools, and shop safety features.

Tools and Materials

Job Sheet 3-2

Drawing board

T-square

Triangles

Pencils

Drawing paper

100-foot tape

Procedure

1. Measure the dimensions of your school's shop and draw the shop outline.
2. Sketch in the stationary tools.
3. Identify the work areas in the shop.
4. Identify the safety features in the shop.
5. Label the power tools.
6. Label the work areas.
7. Label the storage areas.
8. Label the safety features.
9. Submit your drawing to your instructor or share it with your classmates.

Instructor's Score or Approval ____

Job Sheet 3-3

Name ______________________________ Date ____________

Completing the Shop Policies and Procedures Form

Objective

Upon completing this activity, you will have demonstrated your acceptance of the shop safety policy.

Tools and Materials

Job Sheet 3-3

Copy of Figure 3-15 in the textbook

Pen or pencil

Procedure

1. Obtain a copy of the shop safety policy from the instructor.
2. Discuss the policy and procedure form in class.
3. Discuss the form with your parent(s) or guardian.
4. Have the form signed by a parent or guardian.
5. Return the signed form to your instructor.

Instructor's Score or Approval ______________________________

Job Sheet 3-4

Name ______________________________________ Date ______________

Completing the Allergies and Physical Problems Statement

Objective

Upon completing this activity, you will have informed your teacher about any allergic or physical problems that could affect your work in the shop.

Tools and Materials

Job Sheet 3-4

Copy of Figure 3-16 in the textbook

Pen or pencil

Procedure

1. Obtain a copy of Figure 3-16 from the instructor.
2. Discuss the form in class.
3. Complete the form with your parent(s) or guardian.
4. Have the form signed by a parent or guardian.
5. Return the form to your instructor.

Instructor's Score or Approval ____

UNIT 4

Personal Safety in Agricultural Mechanics

Most injuries are avoidable. No one plans to be injured, and most people dislike being reminded of safety precautions; yet injuries occur. In the last unit, you completed a shop policy form with an emphasis on safety. In this unit, you look at specific personal safety items.

Class Activity 4-1: Identifying Safety in the Agricultural Mechanics Shop

Class Activity 4-2: Identifying the Meaning of Shop Color Codes

Shop Activity 4-3: Identifying Shop Color Codes

Class Activity 4-4: Identifying Protective Clothing

Job Sheet 4-1

Name ______________________________ Date ____________

Identifying Safety in the Agricultural Mechanics Shop

Objective

Upon completing this activity, you will have demonstrated the ability to identify safety in the agricultural mechanics shop.

Tools and Materials

Job Sheet 4-1

Pen or pencil

Procedure

Using the textbook as a reference, answer the following questions:

1. Define safety.

____

____

2. How important is it to emphasize safety in the agricultural mechanics shop?

____

____

____

3. What common causes of accidents are listed?

____

____

____

____

____

4. List the precautions needed to provide a safe workplace.

____

____

____

____

____

____

____

____

____

____

5. The best protection from injury is ____________________.
6. The second-best approach to preventing injury is ____________________
____
____.
7. When are goggles or face shields required in the agricultural mechanics shop?
____
____
8. How should long hair be restrained in the shop?
____
____
9. What protective clothing should be used in the shop?
____
____
____
10. What are the characteristics of noise that cause earplugs or earmuffs to be recommended for shop use?
____
____
11. What is the decibel (dB) level OSHA has established as the maximum allowable? __________ dB
12. Why should every shop have an eye wash station?
____
____
____

Instructor's Score or Approval ____________________

Job Sheet 4-2

Name ______________________________ Date ____________

Identifying the Meaning of Shop Color Codes

Objective

Upon completing this activity, you will have demonstrated the ability to identify the meaning of shop color codes.

Tools and Materials

Job Sheet 4-2

Pen or pencil

Procedure

Match the colors in Column I with their meaning in Column II.

Column I	*Column II*
____ 1. Red	a. Warning
____ 2. Orange	b. Mark work areas around objects
____ 3. Yellow	c. Traffic markings
____ 4. Blue	d. Danger
____ 5. Green	e. Caution
____ 6. Black and yellow diagonal stripes	f. Information
____ 7. White	g. Safety
____ 8. White and black stripes	h. Radioactivity
____ 9. Gray	i. Work area markings on shop floors

Instructor's Score or Approval ______________________________

Job Sheet 4-3

Name ______________________________ Date ____________

Identifying Shop Color Codes

Objective

Upon completing this activity, you will have demonstrated the ability to identify the safety color codes.

Tools and Materials

Job Sheet 4-3

A copy of the shop color codes

Pen or pencil

Procedure

Has your agricultural mechanics shop been partially or fully color coded?

____ Yes ____ No

If Yes, proceed; if No, go to Job Sheet 4-4, Identifying Protective Clothing.

1. Safety switches are marked in red. ___ Yes ___ No
2. Fire equipment is marked with red. ___ Yes ___ No
3. Machine hazards are marked in orange. ___ Yes ___ No
4. Backgrounds of switches are orange. ___ Yes ___ No
5. Backgrounds for levers and controls are orange. ___ Yes ___ No
6. Machine controls or adjustments are marked in yellow. ___ Yes ___ No
7. Stationary hazards such as stairs or protruding objects are marked in yellow and black stripes. ___ Yes ___ No
8. Any information signs are in blue. ___ Yes ___ No
9. Traffic areas are marked in white or yellow. ___ Yes ___ No
10. Machine bodies are marked in gray or vista green. ___ Yes ___ No
11. Waste containers are marked in aluminum. ___ Yes ___ No
12. Ivory is used to improve visibility. ___ Yes ___ No

Instructor's Score or Approval ______________________________

Job Sheet 4-4

Name ________________________________ Date ____________

Identifying Protective Clothing

Objective

Upon completing this activity, you will have demonstrated the ability to identify protective clothing.

Tools and Materials

Job Sheet 4-4

Pen or pencil

Procedure

Match the items of protective clothing in Column I with the related items in Column II.

Column I	*Column II*
____ 1. Aprons	a. Hard hat or hairnet
____ 2. Shop coat	b. Decibel
____ 3. Earplugs	c. Bench work
____ 4. Mask	d. Leather or rubber
____ 5. Hair restraint	e. Dust
____ 6. Hard hats	f. Eyes
____ 7. Footwear	g. Frequent changes
____ 8. Coveralls	h. General body protection
____ 9. Gloves	i. Dropped or flying objects
____10. Face shields or goggles	j. Hands

Instructor's Score or Approval ________________________________

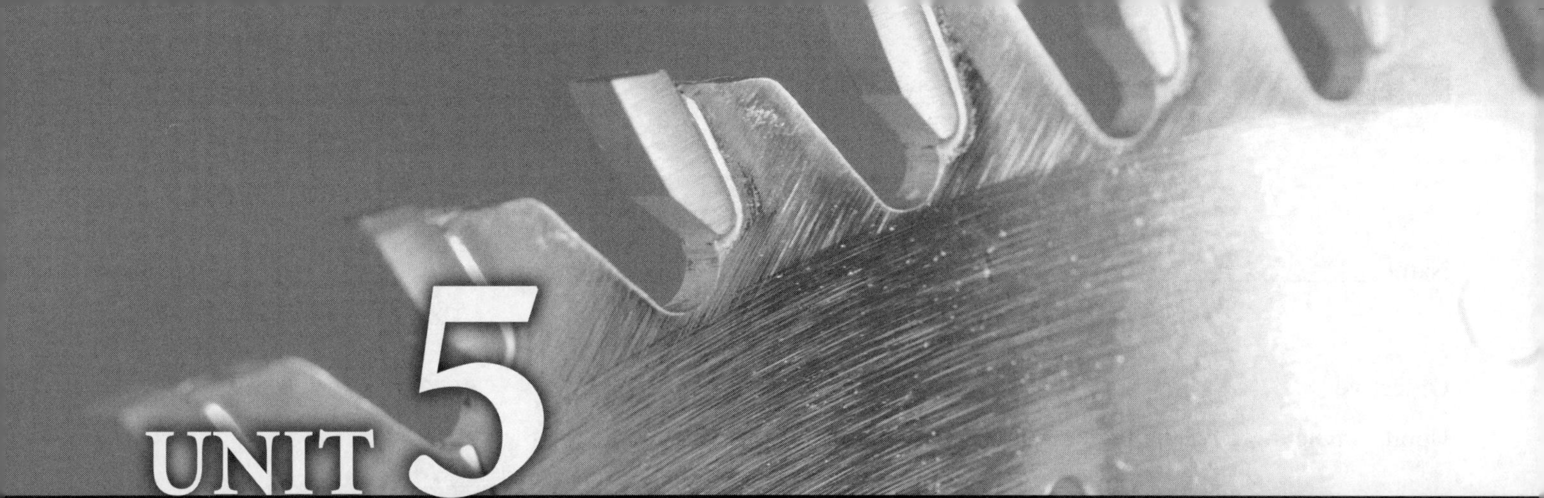

UNIT 5 Reducing Hazards in Agricultural Mechanics

A number of agricultural mechanics activities use processes that may cause a fire if improperly attempted. This unit is geared to helping reduce the fire and other hazards in agricultural mechanics.

Read Unit 5 and refer to it as needed while completing the following activities.

Class Activity 5-1: Identifying Classes of Fires

Class Activity 5-2: Identifying the Components of the Fire Triangle

Class Activity 5-3: Identifying Appropriate Fire Extinguishers

Shop Activity 5-4: Identifying the Types and Locations of Fire Extinguishers in the Agricultural Mechanics Shop

Shop Activity 5-5: Doing a Monthly Fire Extinguisher Checklist

Class Activity 5-6: Identifying the Elements of a Chemical Label

Class Activity 5-7: Identifying the Procedure to Use in Case of an Accident

Class Activity 5-8: Identifying the Procedure to Use in Case of Fire

Class Activity 5-9: Identifying Requirements for the Use of the Slow-Moving Vehicle (SMV) Emblem

Job Sheet 5-1

Name ______________________________________ Date ______________

Identifying Classes of Fires

Objective

Upon completing this activity, you will have demonstrated the ability to identify the classes of fires.

Tools and Materials

Job Sheet 5-1

Pen or pencil

Procedure

Check the materials found in your agricultural mechanics shop in the left column and then label them with the class of fire hazard they represent in the right column.

	Item	*Class*
____	Acetylene	____
____	Aluminum	____
____	Books and manuals	____
____	Electrical equipment	____
____	Gasoline	____
____	Grease	____
____	Lubricating oil	____
____	Lumber	____
____	Oxyfuels	____
____	Paints	____
____	Paper towels	____
____	Plywood	____
____	Protective clothing	____
____	Rope or twine	____
____	Sandpaper	____
____	Varnish	____
____	Welding curtains	____
____	Wooden benches	____

Instructor's Score or Approval ______________________________________

Job Sheet 5-2

Name ______________________________ Date ____________

Identifying the Components of the Fire Triangle

Objective

Upon completing this activity, you will have demonstrated the ability to identify the components of the fire triangle.

Tools and Materials

Job Sheet 5-2

Pen or pencil

Procedure

Answer the following questions:

1. What are the three components of the fire triangle?

   ____
   ____
   ____

2. What happens if one of the three is missing or removed?

   ____
   ____

3. What methods are suggested for removing a part of the fire triangle?

   ____
   ____
   ____

Instructor's Score or Approval ______________________________

Job Sheet 5-3

Name ______________________________ Date ____________

Identifying Appropriate Fire Extinguishers

Objective

Upon completing this activity, you will have demonstrated the ability to identify appropriate fire extinguishers for the fire hazards in the shop.

Tools and Materials

Job Sheet 5-3

Pen or pencil

Procedure

Using the checklist you completed in the first activity, write the class of fire extinguisher to use on a fire hazard by that source in the left column.

Class	*Item*
____	Acetylene
____	Aluminum
____	Books and manuals
____	Electrical equipment
____	Gasoline
____	Grease
____	Lubricating oil
____	Lumber
____	Oxyfuels
____	Paints
____	Paper towels
____	Plywood
____	Protective clothing
____	Rope or twine
____	Sandpaper
____	Varnish
____	Welding curtains
____	Wooden benches

Instructor's Score or Approval ______________________________

Job Sheet 5-4

Name ______________________________ Date ____________

Identifying the Types and Locations of Fire Extinguishers in the Agricultural Mechanics Shop

Objective

Upon completing this activity, you will have demonstrated the ability to identify the types and locations of fire extinguishers in the shop.

Tools and Materials

Job Sheet 5-4

Pen or pencil

Type	*Location*
_____	______________________________
_____	______________________________
_____	______________________________
_____	______________________________
_____	______________________________
_____	______________________________
_____	______________________________
_____	______________________________

Instructor's Score or Approval ______________________________

Job Sheet 5-5

Name ______________________________ Date ____________

Doing a Monthly Fire Extinguisher Checklist

Objective

Upon completing this activity, you will have demonstrated the ability to do a fire extinguisher checklist.

Tools and Materials

Job Sheet 5-5

Pen or pencil

Procedure

1. Obtain the location of the fire extinguisher you are to check from your instructor.
2. Make sure the fire extinguisher is located in the proper area.
3. Be sure the fire extinguisher is the right type for the area.
4. Ensure the fire extinguisher has no visible damage.
5. Make sure safety seals are in place.
6. Make sure the extinguisher is free of dirt and rust.
7. Check that the gauge or indicator is in the operable range.
8. Make sure the extinguisher is of the proper weight.
9. Check that the symbols and nameplate are readable.
10. Make certain there are no nozzle obstructions.
11. When was the unit last serviced? ______________________

Instructor's Score or Approval ______________________________

Job Sheet 5-6

Name ______________________________ Date ____________

Identifying the Elements of a Chemical Label

Objective

Upon completing this activity, you will have demonstrated the ability to identify the elements of a chemical label.

Tools and Materials

Job Sheet 5-6

Chemical label

Pen or pencil

Procedure

Answer the following questions by referencing the chemical label:

1. What is the use classification?

   ____

2. What is the brand name?

   ____

3. What is the common name?

   ____

4. What is the chemical name?

   ____

5. What is the formulation?

   ____

6. What are the ingredients?

   ____

7. What is the signal word?

   ____

8. What is the statement of practical treatment and antidote?

   ____

   ____

   ____

9. What are the directions for use?

   ____

   ____

   ____

10. What is the reentry statement?

    ____

11. What is the precautionary statement?

    ____

    ____

12. What are the storage and disposal directions?

____

____

13. What are the name and address of the manufacturer?

____

____

____

14. What is the EPA registration number?

____

15. What is the EPA establishment number?

____

16. What are the net contents?

____

Instructor's Score or Approval ______________________________

Job Sheet 5-7

Name ______________________________ Date ____________

Identifying the Procedure to Use in Case of an Accident

Objective

Upon completing this activity, you will have demonstrated the ability to identify the procedure to follow in case of an accident.

Tools and Materials

Job Sheet 5-7

Pen or pencil

Procedure

Indicate with a Yes if the following actions should be taken or a No if they should not be taken. Incorrect items are included to test your knowledge of correct procedures.

_____ Avoid unnecessary movement of the victim.

_____ Call or send for help.

_____ Call qualified people for needed help.

_____ Check to see if the victim is breathing.

_____ Clear air passages if necessary.

_____ Control any bleeding.

_____ Cool the victim.

_____ Cover the victim.

_____ Get the victim on his or her feet.

_____ Handle the situation by yourself.

_____ Immobilize the victim if broken bones are suspected.

_____ Keep calm.

_____ Keep the victim lying down.

_____ Move the victim carefully in case of immediate danger such as fire.

_____ Move the victim to a more comfortable location.

_____ Notify the teacher immediately.

_____ Talk to the victim.

_____ Use CPR if necessary.

_____ Wait until help arrives before taking any action.

Have your instructor check your answers and explain any answers that need correction.

Instructor's Score or Approval ______________________________

Job Sheet 5-8

Name ______________________________ Date ____________

Identifying the Procedure to Use in Case of Fire

Objective

Upon completing this activity, you will have demonstrated the ability to identify the procedure to use in case of a fire in the shop.

Tools and Materials

Job Sheet 5-8

Pen or pencil

Procedure

Write Yes or No to indicate which actions to take in case of a fire.

_____ Avoid causing others to panic.

_____ Call everyone over to the fire scene.

_____ Clear the area.

_____ Call the fire department.

_____ Carry tools out of the shop.

_____ Notify the teacher.

_____ Scream at everyone.

_____ Set off the fire alarm.

_____ Try to keep others calm.

_____ Use any fire extinguisher on hand.

_____ Use the appropriate fire extinguisher if it can extinguish the fire.

Have your instructor check your answers and explain to you any answers that need correction.

Instructor's Score or Approval ______________________________

Job Sheet 5-9

Name ______________________________ Date ____________

Identifying Requirements for the Use of the Slow-Moving Vehicle (SMV) Emblem

Objective

Upon completing this activity, you will have demonstrated your ability to identify the requirements for the SMV emblem's use.

Tools and Materials

Job Sheet 5-9

Pen or pencil

Procedure

Check the items that apply to the slow-moving vehicle emblem requirements.

_____ Any vehicle that travels less than 25 miles per hour on public highways

_____ Reflective orange triangle with a red border

_____ Is displayed at the front of the vehicle

_____ Is displayed on the rear of the vehicle

_____ Does not apply to towed wagons

_____ Applies to any farm machine towed at 25 mph or less

_____ Faded or damaged emblems should be replaced

Instructor's Score or Approval ______________________________

UNIT 6
Shop Cleanup and Organization

When many people use the same facility and tools, it is obvious that some order is essential so each person can find the tools necessary for their activities. It is also apparent that safety is important, and a clean, well-organized workplace promotes safety.

Teamwork and cooperation are needed in shop cleanup. Each student must be responsible for the work area used by that student and must also participate in the general cleanup of the whole area.

Read Unit 6 and refer to it as needed while completing the following activities.

Shop Activity 6-1: Identifying the Shop Organization Aids

Shop Activity 6-2: Identifying a Suitable Shop Cleanup Plan

Job Sheet 6-1

Name ______________________________ Date ____________

Identifying the Shop Organization Aids

Objective

Upon completing this activity, you will have demonstrated the ability to identify the shop organization aids.

Tools and Materials

Job Sheet 6-1

Pen or pencil

Procedure

Check the shop organization aids that are present in the shop.

____ Broom racks

____ Dust collection system

____ Fenced outside storage area

____ Locked tool panels

____ Lumber storage racks

____ Metal cans for rags

____ Metal storage for paints and materials

____ Metal storage racks

____ Paint room or booth

____ Parts cleaner

____ Project storage area

____ Scrap wood container

____ Shop vacuum

____ Shop cleanup chart or sheet

____ Tool storage room

____ Trash containers

____ Waste metal storage bin or containers

____ Other:

____ Other:

Instructor's Score or Approval ______________________________

Job Sheet 6-2

Name ______________________________ Date ____________

Identifying a Suitable Shop Cleanup Plan

Objective

Upon completing this activity, you will have demonstrated the ability to identify a suitable cleanup plan.

Tools and Materials

Job Sheet 6-2

Pen or pencil

Procedure

Check the elements present in your shop cleanup plan.

____ All students participate.

____ All students share responsibility.

____ All tools are put away and accounted for.

____ Benches are cleaned.

____ Cabinets and storage areas are locked.

____ Cleaning items are properly stored.

____ Cleanup starts on signal.

____ Floors are cleaned.

____ Jobs are checked for completion.

____ Job responsibilities rotate.

____ Machines are vacuumed or cleaned.

____ Projects are properly stored.

____ Students are evaluated.

____ Students wait for instructor's dismissal.

____ Sinks and restrooms are clean and orderly.

____ Supplies are properly stored.

____ Other: ______________________

____ Other: ______________________

Discuss these elements with your instructor to see if your cleanup plan should be modified.

List any elements that should be added to your shop cleanup plan.

____

Instructor's Score or Approval ______________________________

SECTION 3

Hand Woodworking and Metalworking

Unit 7
Hand Tools, Fasteners, and Hardware

Unit 8
Layout Tools and Procedures

Unit 9
Selecting, Cutting, and Shaping Wood

Unit 10
Fastening Wood

Unit 11
Finishing Wood

Unit 12
Identifying, Marking, Cutting, and Bending Metal

Unit 13
Fastening Metal

UNIT 7

Hand Tools, Fasteners, and Hardware

A basic knowledge of hand tools and their use is essential to most mechanical jobs because some portion of each job will require their use. Thus, you should be able to identify the common shop tools and understand their uses. You should also learn to use hand tools before moving on to power tool operation.

Class Activity 7-1: Identifying the Classes of Tools

Class Activity 7-2: Classifying Tools

Class Activity 7-3: Identifying Common Fasteners

Class Activity 7-4: Identifying Common Hardware

Job Sheet 7-1

Name ________________________________ Date ______________

Identifying the Classes of Tools

Objective

Upon completing this activity, you will be able to determine the classification of tools by their use.

Tools and Materials

Job Sheet 7-1

Pen or pencil

Procedure

Match the uses of the tools with their classifications.

____	1. Layout tools	a. Tools used to move another tool using force
____	2. Saws	b. Tools and extensions used to turn nuts or bolts and screws
____	3. Boring tools	c. Tools used to measure or mark materials
____	4. Hammers	d. Tools used to loosen or remove earth
____	5. Pliers	e. Tools used to make holes or change their shapes
____	6. Wrenches	f. Tools used to hold materials for the use of other tools
____	7. Digging tools	g. Tools used to cut or remove material
____	8. Chisels and punches	h. Tools used to turn slot-headed screws or bolts
____	9. Screwdrivers	i. Tools used to draw and hold objects together
____	10. Clamps	j. Tools designed to take structures apart
____	11. Wrecking tools	k. Tools usually struck by a hammer

Instructor's Score or Approval ________________________________

Job Sheet 7-2

Name ______________________________ Date ____________

Classifying Tools

Objective

Upon completing this activity, you will have demonstrated the ability to identify the classification of a tool.

Tools and Materials

Job Sheet 7-2

Pen or pencil

Procedure

List the correct category (Boring, Chisel/Punch, Clamp, Digging, Hammer, Layout, Pliers, Saw, Screwdriver, Wrecking, or Wrench) next to each tool listed.

____ 1. Adjustable wrench

____ 2. Axe

____ 3. Backsaw

____ 4. Ball-peen hammer

____ 5. Bar clamp

____ 6. Bench vise

____ 7. Bevel

____ 8. Blacksmith's hammer

____ 9. Block clamp

____ 10. Bolt cutters

____ 11. Bow saw

____ 12. Caliper rule

____ 13. Cape chisel

____ 14. Carpenter's square

____ 15. C-clamp

____ 16. Center punch

____ 17. Chalk line

____ 18. Cold chisel

____ 19. Combination square

____ 20. Compass saw

____ 21. Compound snips

____ 22. Concrete edger

____ 23. Concrete groover

____ 24. Coping saw

____ 25. Cordless power screwdriver

____ 26. Corner clamp

____ 27. Cultivator

____ 28. Countersink

____ 29. Curved-claw hammer

____ 30. Cutter mattock

____ 31. Deep socket

____ 32. Diagonal cutters

____ 33. Diamond-point chisel

____ 34. Die

____ 35. Diestock

____ 36. Dividers

____ 37. Drive extension

____ 38. Fencing tool

____ 39. Finishing trowel

____ 40. Flat file

____ 41. Flex handle

____ 42. Flex socket

____ 43. Folding wood rule

____ 44. Garden hoe

____ 45. Garden rake
____ 46. Glass cutter
____ 47. Grease gun
____ 48. Groove joint pliers
____ 49. Hacksaw
____ 50. Handsaw, crosscut
____ 51. Handsaw, rip
____ 52. Hand trowel
____ 53. Half-round file
____ 54. Hatchet
____ 55. Hex key
____ 56. Hole saw
____ 57. Inside caliper
____ 58. Leaf gauge
____ 59. Level
____ 60. Lever wrench pliers
____ 61. Line level
____ 62. Long-handled shovel
____ 63. Long-handled spade
____ 64. Long-nose pliers
____ 65. Machinist's vise
____ 66. Marking gauge
____ 67. Masonry drill
____ 68. Mason's hammer
____ 69. Mini hacksaw
____ 70. Nippers
____ 71. Offset box wrench
____ 72. Open-end wrench
____ 73. Outside caliper
____ 74. Pipe cutters
____ 75. Pipe wrench
____ 76. Phillips screwdriver bit
____ 77. Phillips screwdriver
____ 78. Pick mattock
____ 79. Pin punch
____ 80. Pipe vise
____ 81. Plastic mallet
____ 82. Plug cutter
____ 83. Pointing trowel
____ 84. Posthole digger
____ 85. Power bit
____ 86. Power drill
____ 87. Prick punch
____ 88. Pruning saw
____ 89. Pruning shears
____ 90. Push drill
____ 91. Pry bar
____ 92. Ratchet offset screwdriver
____ 93. Ratchet handle
____ 94. Ripping chisel
____ 95. Ripping claw hammer
____ 96. Round file
____ 97. Rubber mallet
____ 98. Screwmate
____ 99. Scratch awl
____ 100. Scraper
____ 101. Sledge hammer
____ 102. Sliding T handle
____ 103. Slip joint pliers
____ 104. Socket adapter
____ 105. Spading fork
____ 106. Speed handle
____ 107. Splitting wedge
____ 108. Spring clamp
____ 109. Standard screwdriver bit
____ 110. Standard screwdriver
____ 111. Starter punch
____ 112. Steel tape
____ 113. Surform
____ 114. Tap

____ 115. Tap wrench

____ 116. Tin snips

____ 117. Torque wrench

____ 118. Triangular file

____ 119. Try square

____ 120. Turf edger

____ 121. Twist drill

____ 122. Utility knife

____ 123. Universal joint

____ 124. Wire gauge

____ 125. Wood chisel

____ 126. Wrecking bar

Instructor's Score or Approval ____________________________________

Job Sheet 7-3

Name ______________________________ Date ____________

Identifying Common Fasteners

Objective

Upon completing this activity, you will have demonstrated the ability to identify the common fasteners used in agricultural mechanics.

Tools and Materials

Job Sheet 7-3

Pen or pencil

Procedure

Using the textbook, identify and label each of the common fasteners.

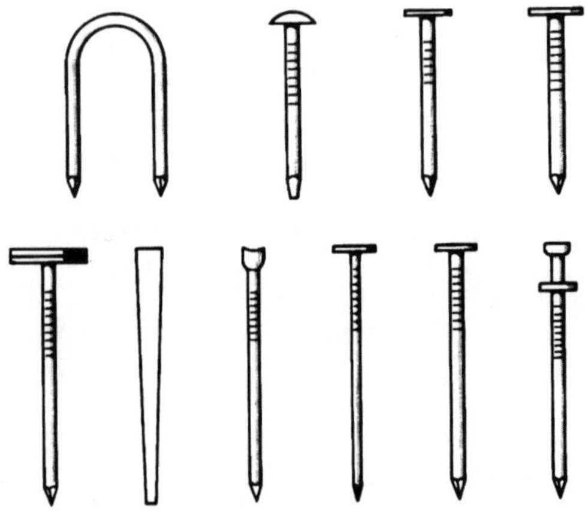

Figure 7-1 Types of nails

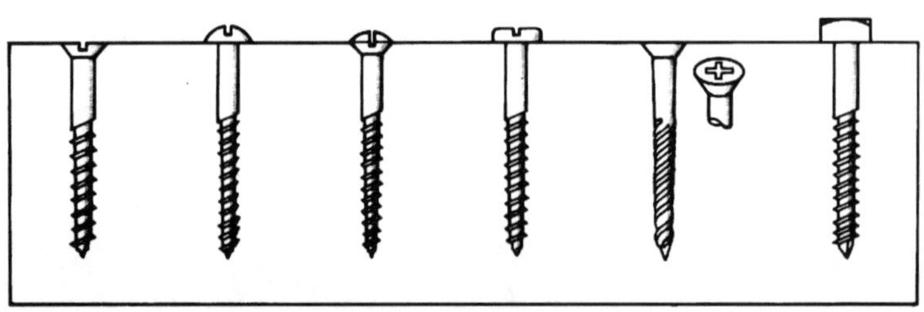

Figure 7-2 Screw classifications

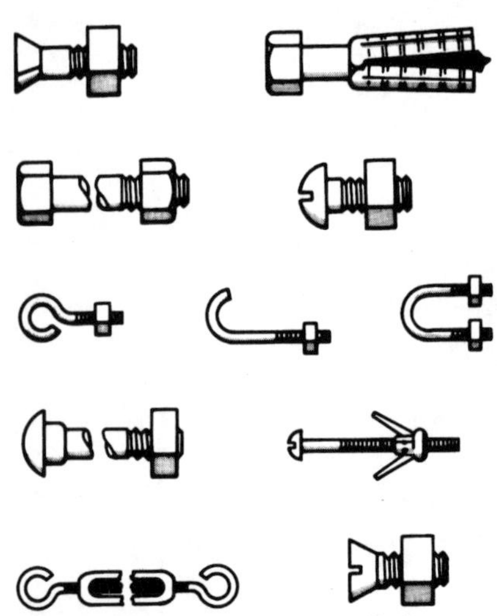

Figure 7-3 Common bolts

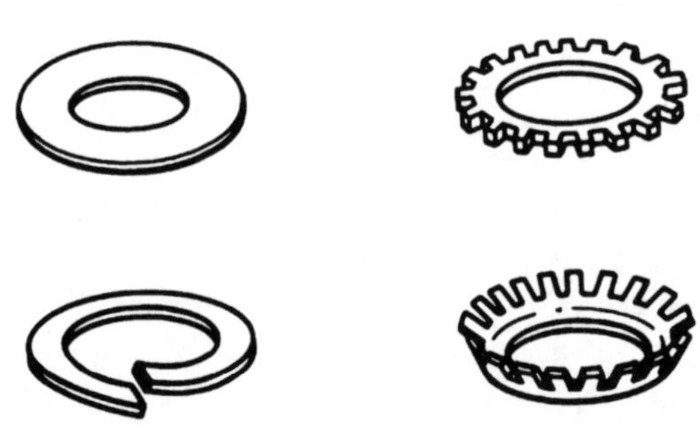

Figure 7-4 Types of washers

Instructor's Score or Approval ______________________________

Job Sheet 7-4

Name ______________________________ Date ______________

Identifying Common Hardware

Objective

Upon completing this activity, you will have demonstrated the ability to identify the common hardware items.

Tools and Materials

Job Sheet 7-4

Pen or pencil

Procedure

Using the textbook, identify and label the common hardware items.

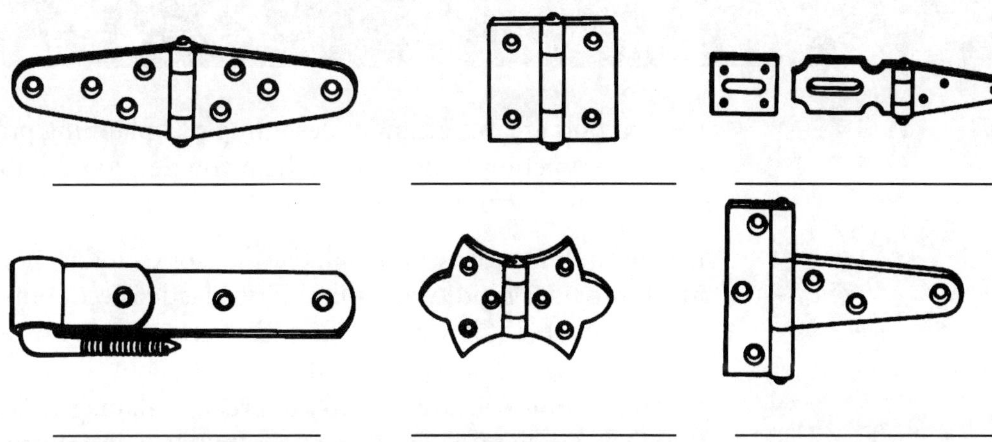

Figure 7-5 Types of hinges

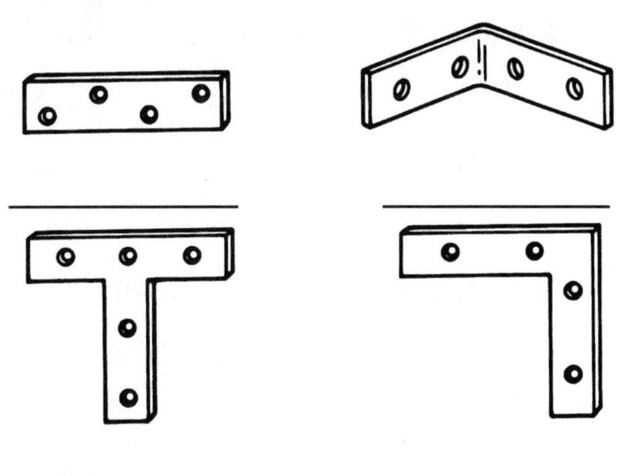

Figure 7-6 Flush plates

Instructor's Score or Approval ______________________________

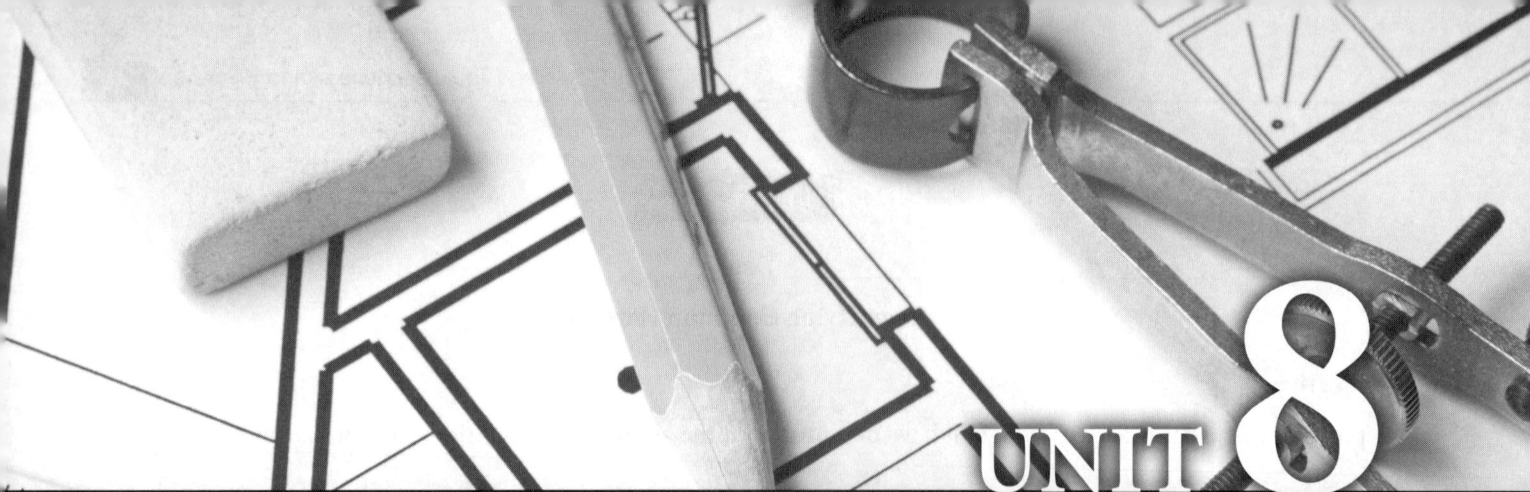

Layout Tools and Procedures

Layout tools are important in designing and planning projects for shop construction. This unit will help you develop the ability to properly use layout tools.

Layout tools are made of metal, plastic, wood, and sometimes cloth. Metal provides rigidity and durability. Plastic is economical and lightweight, but it is less durable and easily damaged by heat or chemicals. Wood is usually used for handles but is not typically used for layout tools because it is easy to break or damage. Cloth is sometimes used in longer measuring tapes to lessen weight and bulk.

Class Activity 8-1: Identifying Inch and Metric Ruler Markings

Class Activity 8-2: Identifying Layout Tools

Class Activity 8-3: Identifying the Uses of Layout Tools

Shop Activity 8-4: Using a Square

Shop Activity 8-5: Squaring a Large Area

Shop Activity 8-6: Leveling an Object

Shop Activity 8-7: Creating a Pattern

Job Sheet 8-1

Name ______________________________ Date __________

Identifying Inch and Metric Ruler Markings

Objective

Upon completing this activity, you will have demonstrated the ability to identify the markings on inch and metric rulers.

Tools and Materials

Job Sheet 8-1

Pen or pencil

Procedure

Fill in the blanks on the accompanying figures with the correct inch or metric readings.

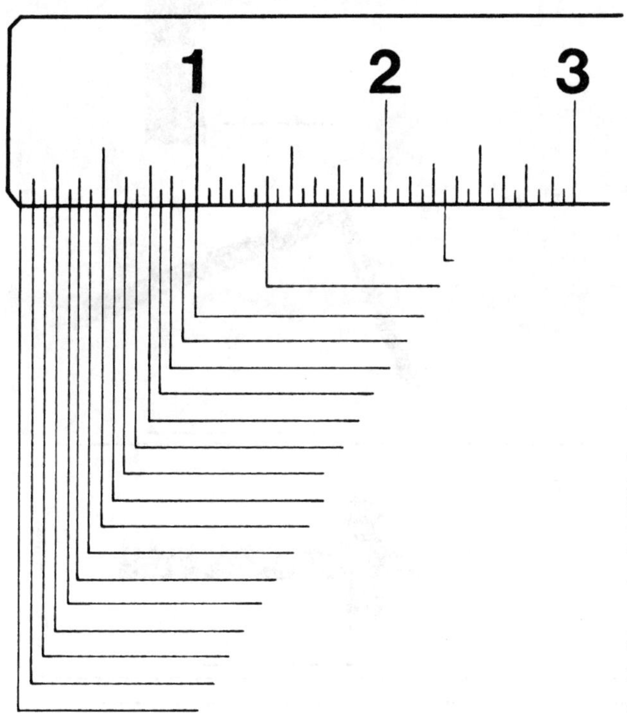

Figure 8-1 Inch ruler readings

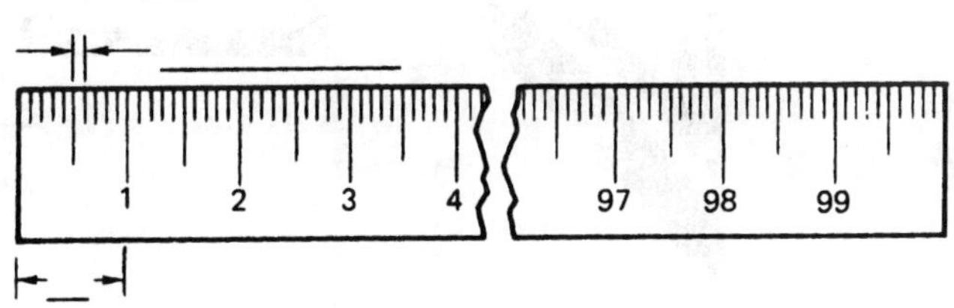

Figure 8-2 Metric ruler readings

Instructor's Score or Approval ______________________________

Job Sheet 8-2

Name ______________________________ Date ____________

Identifying Layout Tools

Objective

Upon completing this activity, you will have demonstrated the ability to identify layout tools.

Tools and Materials

Job Sheet 8-2

Pen or pencil

Procedure

Using the accompanying figures, write in the correct names for the layout tools pictured.

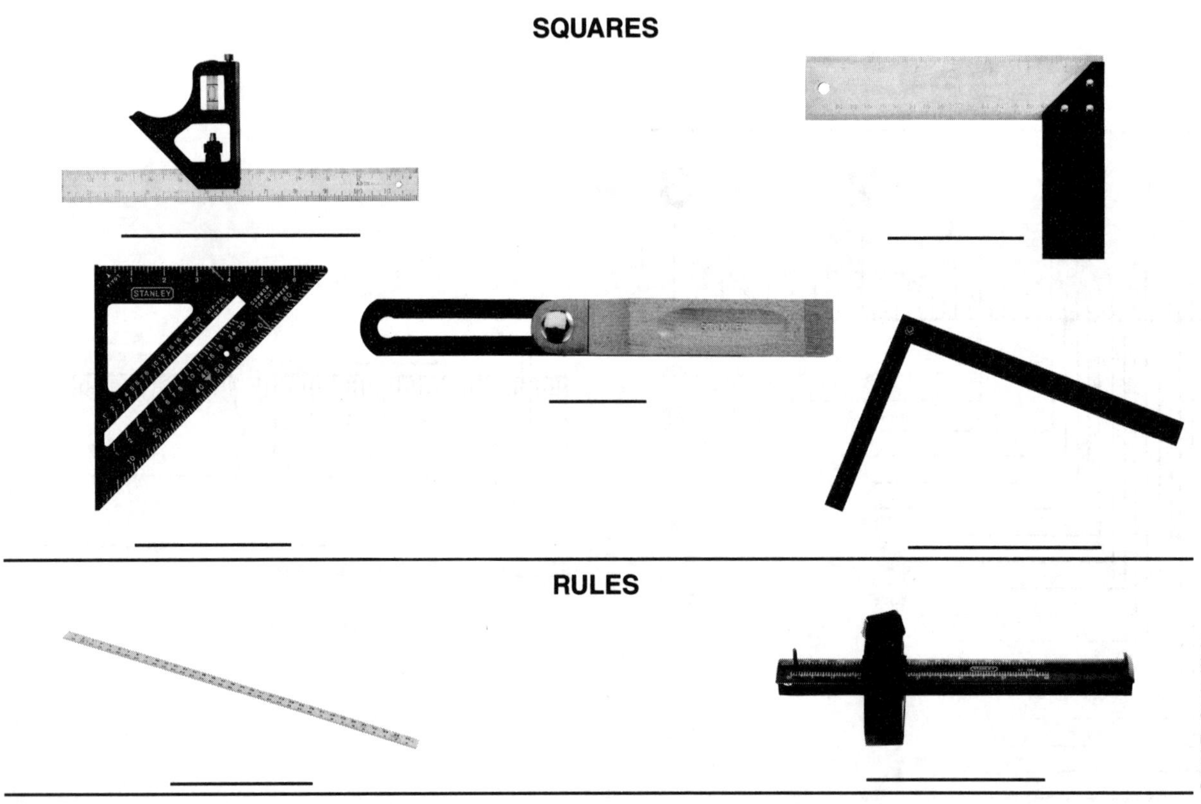

Figure 8-3 Layout tools (Continued)

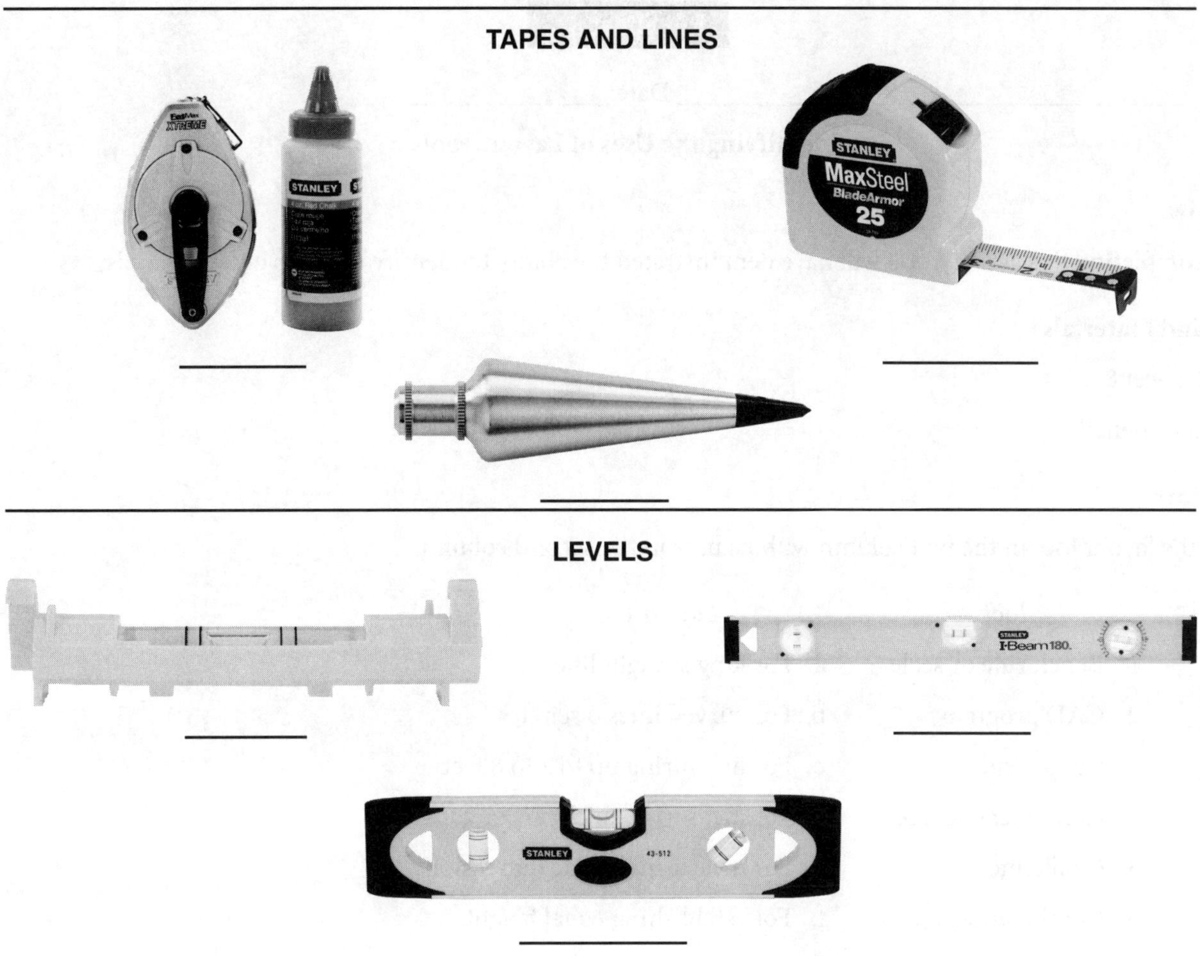

Figure 8-3 (Continued)

Instructor's Score or Approval ______________________________

Job Sheet 8-3

Name ______________________________ Date ______________

Identifying the Uses of Layout Tools

Objective

Upon completing this activity, you will have demonstrated the ability to identify the uses of layout tools.

Tools and Materials

Job Sheet 8-3

Pen or pencil

Procedure

Match the layout tool in the first column with its uses in the second column.

	Tool	*Layout Use*
____	1. Bench rule or scale	a. For long straight lines
____	2. CAD program	b. For curved lines or circles
____	3. Caliper rule	c. For measuring up to 2 to 8 feet
____	4. Carpenter's square	d. For measuring thickness or gaps
____	5. Chalk line	e. For measuring inside dimensions
____	6. Combination square	f. For establishing equal heights
____	7. Dividers and scribers	g. Guide for laying blocks
____	8. Folding rule	h. For very accurate small measuring
____	9. Gauges	i. For measuring outside diameters
____	10. Inside caliper	j. Computer-assisted design
____	11. Laser level	k. For transferring a vertical point
____	12. Level or line level	l. Construction framing
____	13. Line	m. Rigid measuring tool
____	14. Outside caliper	n. For transferring or marking angles
____	15. Pattern	o. Projects a level beam of light
____	16. Plumb bob	p. For copying objects
____	17. Try square	q. For testing accuracy of cuts
____	18. Sliding T bevel	r. Flexible measuring rule
____	19. Tape rule	s. Versatile measuring or testing tool

Instructor's Score or Approval ______________________________

Job Sheet 8-4

Name ______________________________ Date ____________

Using a Square

Objective

Upon completing this activity, you will have demonstrated the ability to use a square to mark a board and to check the accuracy of a saw cut.

Tools and Materials

Job Sheet 8-4

Combination or try square

Handsaw

Sawhorse

Board

Pencil

Procedure

1. Place the board flat on the sawhorse.
2. Place the square on the board ½" from the end.
3. Mark along the blade of the square.
4. Turn the board on its edge and place the blade of the square at the start of the first mark.
5. Mark the edge of the board along the blade of the square.
6. With the board and marks extending over the end of the sawhorse, make the saw cut.
7. Use the square to check the accuracy of the cut.

Instructor's Score or Approval ______________________________

Job Sheet 8-5

Name ____________________________________ Date ______________

Squaring a Large Area

Objective

Upon completing this activity, you will have demonstrated the ability to square a large area.

Tools and Materials

Job Sheet 8-5

Carpenter's square

50' tape

String

Stakes

Procedure

1. Set a stake to establish a first corner.
2. Place the heel of the carpenter's square against the stake.
3. Attach a string and stretch it along one leg of the carpenter's square. Stake the other end at 20'.
4. Attach a second string and stretch it along the other leg at a right angle. Stake the other end at 15'.
5. Measure 6' from the corner on the second side and tie a piece of string at that point.
6. Measure 8' from the corner on the second side and tie another piece of string at that point.
7. Measure between the two pieces of string.
8. Move *one* string in or out until the distance is exactly 10' (+ or – 1/16").
9. Place the heel of the carpenter's square at the third stake with one leg against the second string and the other pointing in a direction parallel to the second stake.
10. Stretch a string along the leg of the carpenter's square and stake the other end at 20'.
11. Repeat Steps 5 through 8.
12. Run a string from the fourth stake back to the second stake to complete the rectangle. This distance should measure 15'.
13. Measure the diagonals and adjust the corners until they are equal.

Instructor's Score or Approval ____________________________________

Job Sheet 8-6

Name ______________________________ Date ____________

Leveling an Object

Objective

Upon completing this activity, you will have demonstrated the ability to level an object.

Tools and Materials

Job Sheet 8-6

Level

Sawhorse

Shims

An uneven surface

Procedure

1. Set the sawhorse on the uneven surface.
2. Set the level on the sawhorse.
3. Place shims under the legs that are farthest away from the bubble in the level. This will start to center the bubble.
4. Repeat this process on the other legs as necessary until the bubble is centered in the level.
5. Return the sawhorse and level.

Instructor's Score or Approval ______________________________

Job Sheet 8-7

Name ______________________________ Date ____________

Creating a Pattern

Objective

Upon completing this activity, you will have demonstrated the ability to create a pattern.

Tools and Materials

Job Sheet 8-7

Figure 8-26 from the textbook

8' × 10" tagboard

Ruler

Pencil

Procedure

1. Fold the tagboard in half lengthwise.
2. Make a grid with 1" spacing.
3. Draw the tree as shown in Figure 8-26 in the textbook on the enlarged grid.
4. Cut out the tree.
5. Unfold the tree.

Instructor's Score or Approval ______________________________

UNIT 9

Selecting, Cutting, and Shaping Wood

Lumber is the most common building material today because of its low cost and workability. You should become familiar with its characteristics and its shaping.

Although power saws have usually replaced handsaw use, the principles and practices used in these exercises will be useful when you later learn the use of power tools.

Shop Activity 9-1: Identifying Common Lumber Species

Shop Activity 9-2: Identifying Nominal and Actual Lumber Dimensions

Shop Activity 9-3: Identifying Handsaws

Shop Activity 9-4: Using a Crosscut Handsaw

Shop Activity 9-5: Using a Ripsaw

Shop Activity 9-6: Using a Surform to Smooth an Edge

Shop Activity 9-7: Making a Dado

Job Sheet 9-1

Name ______________________ Date __________

Identifying Common Lumber Species

Objective

Upon completing this activity, you will have demonstrated the ability to identify common species of lumber.

Tools and Materials

Lumber samples of common species

Job Sheet 9-1

Pencil or pen

Procedure

Take the samples of the common species provided by your instructor and complete the following information table by observation. Use as many species as your instructor has available. Study the samples until you can correctly identify other samples supplied without identifications. Complete the remainder of the table by using the textbook.

Wood	*Hard*	*Medium*	*Soft*	*Color*
Birch	____	____	____	____
Cedar, Red	____	____	____	____
Cherry	____	____	____	____
Cypress	____	____	____	____
Fir	____	____	____	____
Hemlock	____	____	____	____
Maple	____	____	____	____
Mahogany	____	____	____	____
Oak	____	____	____	____
Pine, Yellow	____	____	____	____
Pine, White	____	____	____	____
Redwood	____	____	____	____
Walnut, Black	____	____	____	____
Willow, Black	____	____	____	____
____	____	____	____	____
____	____	____	____	____
____	____	____	____	____

Instructor's Score or Approval ______________________

Job Sheet 9-2

Name ______________________________ Date ____________

Identifying Nominal and Actual Lumber Dimensions

Objective

Upon completing this activity, you will have demonstrated the ability to identify the actual sizes of nominal dimensions of lumber.

Tools and Materials

Job Sheet 9-2

Tape, ruler, or carpenter's square

Samples of the following lumber items: 1" × 2", 1" × 4", 1" × 6", 1" × 8", 1" × 10", 1" × 12", 2" × 2", 2" × 4", 2" × 6", 2" × 8", 2" × 10", 2" × 12", 4" × 4", and 4" × 6"

Pen or pencil

Procedure

Using the measuring rule, complete the following table:

Nominal Size	*Actual Size Measured*
1" × 2"	____
1" × 4"	____
1" × 6"	____
1" × 8"	____
1" × 10"	____
1" × 12"	____
2" × 2"	____
2" × 4"	____
2" × 6"	____
2" × 8"	____
2" × 10"	____
2" × 12"	____
4" × 4"	____
4" × 6"	____

Now that you've measured the actual sizes of these pieces of lumber, what rules or generalizations could you make about the differences in nominal size and actual size?

Instructor's Score or Approval ______________________________

Job Sheet 9-3

Name ______________________________ Date ______________

Identifying Handsaws

Objective

Upon completing this activity, you will have demonstrated the ability to identify handsaws.

Tools and Materials

Job Sheet 9-3

Clipboard

Pen or pencil

Procedure

Take the job sheet to the shop and complete it by correctly identifying the handsaws in the tool room or cabinets.

Number	*Saws*	*Tooth Points Per Inch*	*Location*
___	Crosscut	___	___
___	Rip	___	___
___	Keyhole	___	___
___	Compass	___	___
___	Back	___	___

Instructor's Score or Approval ______________________________

Job Sheet 9-4

Name ____ Date ______________

Using a Crosscut Handsaw

Objective

Upon completing this activity, you will have demonstrated the ability to properly use a crosscut saw.

Tools and Materials

- Job Sheet 9-4
- 1" × 6" × 18" board
- Crosscut handsaw
- Pencil
- Try square
- Sawhorse

Procedure

1. Place the 1" × 6" on the sawhorse.
2. Place the try square against the edge of the board and mark a line ½" from the end.
3. Turn the 1" × 6" on its edge and mark a line with the square touching the start of the first line.
4. Extend the board over the end of the sawhorse so the line is beyond the sawhorse.
5. Hold the board on the sawhorse with your left knee.*
6. Hold the saw with your index finger along the handle.
7. Align your right shoulder, knee, elbow, and eye with the line.**
8. Place your thumbnail on the mark to aid the start of the cut.
9. Place the saw on the mark at a 45° angle so you see the back of the saw and the line aligned.
10. Start the cut with a backstroke.
11. Saw the board with long even strokes. Do not put pressure on the last stroke.
12. Check the board with the try square for accuracy and squareness.
13. Repeat these steps until you produce square cuts.
14. Clean the area and put away the tools.

Instructor's Score or Approval ____

*Use your right knee if you are left-handed.

**Use your left shoulder, knee, elbow, and eye if you are left-handed.

Job Sheet 9-5

Name ______________________________ Date ____________

Using a Ripsaw

Objective

Upon completing this activity, you will have demonstrated the ability to use a ripsaw.

Tools and Materials

Job Sheet 9-5

The board used in Activity 9-4

Ripsaw

Sawhorse

Combination square

Pencil

Procedure

1. Place the board on the sawhorse.
2. Set the combination square at 4⅛".
3. Place the pencil at the end of the square and draw a line the length of the board.
4. Rip the board using the technique from Activity 9-4.
5. Check the board for a square cut.
6. Clean the area and put the tools away.

Instructor's Score or Approval ______________________________

Job Sheet 9-6

Name ______________________________ Date ____________

Using a Surform to Smooth an Edge

Objective

Upon completing this activity, you will have demonstrated the ability to use a Surform to smooth an edge.

Tools and Materials

Job Sheet 9-6

The board used in previous activities

Surform

Combination square

Bench vise

Pencil

Procedure

1. Set the combination square at 4".
2. Mark a line at 4" down the length of the board on the same side as the rip cut made in the activity on Job Sheet 9-5.
3. Mark a second 4" line down the length of the board on the underside.
4. Place the board in the vise so both lines are above the rise and run horizontally.
5. Push the Surform at an angle with the grain using long smooth strokes.
6. Check the edge for squareness.
7. Remove the additional wood above the line until the edge is straight, smooth, and square.
8. Clean the area and put away the tools.

Instructor's Score or Approval ______________________________

Job Sheet 9-7

Name ____ Date _______________

Making a Dado

Objective

Upon completing this activity, you will have demonstrated the ability to make a dado.

Tools and Materials

Job Sheet 9-7

Crosscut saw

Two pieces of 1" × 6" lumber

½" wood chisel

Combination square

Bench vise

Pencil

Procedure

1. Use the square to draw two parallel lines across one of the boards exactly ¾" apart.
2. Use the square to extend these two lines across both edges of the board.
3. Set the combination square at ⅜" and mark the depth of cut on each edge of the board.
4. Place the board in the bench vise.
5. Saw a kerf across the board on each line to the depth of cut.
6. Saw additional kerfs between the two lines to the depth of cut.
7. Remove the wood between the kerfs with the wood chisel.
8. Fit the second board into the dado.
9. Check for a 90° angle between the two boards.
10. Remove more wood, if needed, to perfect the fit.
11. Clean the area and put away the tools.

Instructor's Score or Approval ____

UNIT 10

Fastening Wood

Wood is very important in agriculture for the construction of buildings, fences, parts of tools, and machinery. The ability to fasten wooden parts with strong joints is a valuable skill. This unit is concerned with the types of joints and fasteners that keep the joints rigid and secure.

Class Activity 10-1: Identifying Common Wood Joints

Class Activity 10-2: Understanding Nailing Techniques

Shop Activity 10-3: Fastening Wood with Nails

Shop Activity 10-4: Fastening Wood with Screws

Shop Activity 10-5: Fastening Wood with Bolts

Shop Activity 10-6: Fastening Wood with Glue and Screws

Shop Activity 10-7: Fastening Wood with Dowels and Glue

Job Sheet 10-1

Name ______________________________________ Date ______________

Identifying Common Wood Joints

Objective

Upon completing this activity, you will have demonstrated the ability to identify the common wood joints by description or pictures.

Tools and Materials

Job Sheet 10-1

Pen or pencil

Procedure

Match the description of the joint with the name of the joint, and then label the pictures of the joints correctly.

____	1. Butt joint	a. A joint made by interlocking parts of two pieces
____	2. Lap joint	b. A joint made by placing two pieces end to end or side to side
____	3. Dado and rabbet joint	c. A joint made by cutting the ends of two pieces at 45°, then joining at 90°
____	4. Miter joint	d. A joint formed by fastening one piece face to face with another
____	5. Dovetail joint	e. A joint made by making a rectangular groove in a board or at the end to receive the end of the second piece

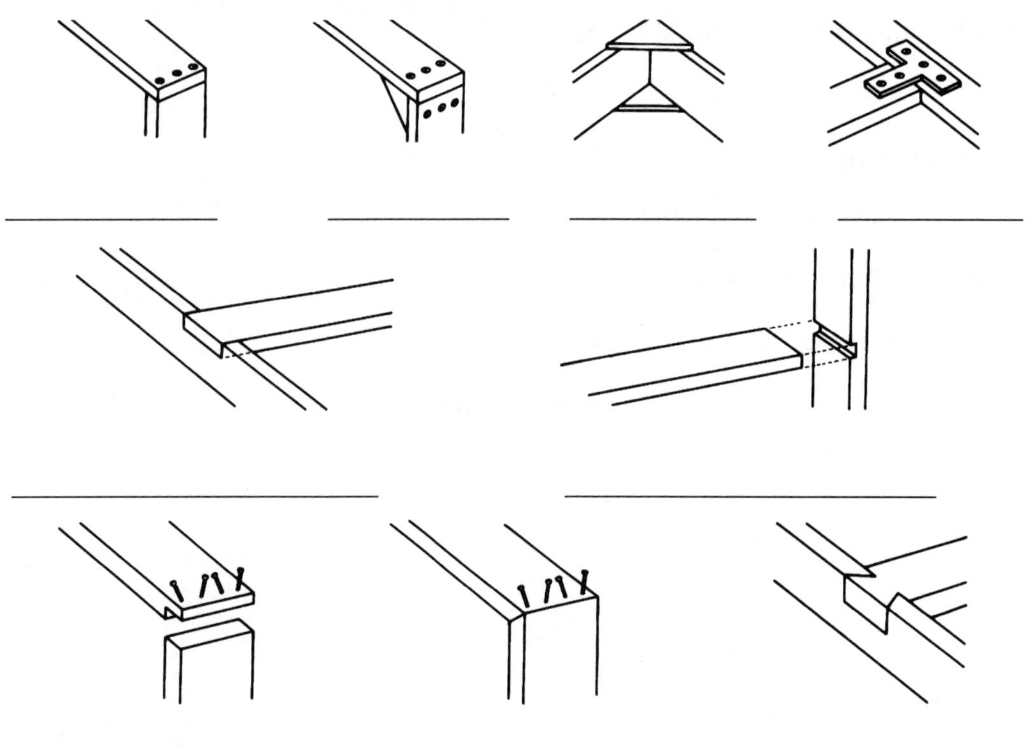

Figure 10-1 Common wood joints

Instructor's Score or Approval ______________________________________

Job Sheet 10-2

Name ______________________________ Date ____________

Understanding Nailing Techniques

Objective

Upon completing this activity, you will have demonstrated an understanding of nailing techniques.

Tools and Materials

Job Sheet 10-2

Pen or pencil

Procedure

Answer the following questions:

1. Why is a block of wood recommended when pulling nails?
___
___
2. When performing finish work, what is the purpose of a nail set?
___
___
3. What is toe nailing and when is it used?
___
___
___
4. What is end nailing and when is it used?
___
___
___
5. What is clinching and why is it used?
___
___
6. What are staples and how are they used?
___
___
___
___
___

Instructor's Score or Approval ______________________________

Job Sheet 10-3

Name ____ Date _______________

Fastening Wood with Nails

Objective

Upon completing this activity, you will have demonstrated the ability to fasten wood with nails.

Tools and Materials

Job Sheet 10-3

Two pieces of 2" × 4" wood

Curved-claw hammer

8d common nails

Procedure

1. Place the first piece of wood across the second piece so the two ends are even with the sides (this should resemble an "L" shape).
2. Select a point for the first nail by staying the width of the board away from the edge of the board.
3. Start a nail by holding it perpendicular to the board and tapping it lightly.
4. Drive the nail while holding the hammer by the handle.
5. Use the wrist and arm to deliver firm blows.
6. Have the face of the hammer parallel to the board when the blow is struck.
7. Drive the nail until the head is flush with the wood and the two boards are securely fastened.
8. Remove the nails and put away the tools.

Instructor's Score or Approval ____

Job Sheet 10-4

Name ______________________________ Date ____________

Fastening Wood with Screws

Objective

Upon completing this activity, you will have demonstrated the ability to fasten wood with screws.

Tools and Materials

Job Sheet 10-4

A piece of 1" × 4"

A piece of 2" × 4"

Standard screwdriver

Four No. 8 × 2" flat-head screws

Countersink

Shank hole drill

Pilot hole drill

Power drill

Procedure

1. Place the 1" × 4" across the end of the 2" × 4" so the end of each board is even with the edge of the other board.
2. Mark four locations for the screws.
3. Select the proper shank hole drill (see Figure 10-8 in the textbook).
4. Install the drill in the power drill.
5. Drill the shank holes.
6. Mark the location of the pilot holes.
7. Install the pilot hole drill.
8. Drill the pilot holes 1" to 1¼" deep.
9. Install the countersink in the drill.
10. Countersink the shank holes so the screws will be flush when tightened.
11. Install the screws and turn them until the heads are flush.
12. Clean the area and put away the tools.

Instructor's Score or Approval ______________________________

Job Sheet 10-5

Name ______________________________ Date ____________

Fastening Wood with Bolts

Objective

Upon completing this activity, you will have demonstrated the ability to fasten wood with bolts.

Tools and Materials

Job Sheet 10-5

Two $\frac{5}{16}$" × 3½" carriage bolts

Two 2" × 4" boards

Power drill

A $\frac{5}{16}$" drill bit

Two $\frac{5}{16}$" washers and nuts

Bench vise

½" wrench

Procedure

1. Mark two locations for the bolts.
2. Clamp one board in the vise to drill the holes.
3. Insert the drill bit in the power drill.
4. Drill the first 2" × 4", remembering to complete each hole from the back side.
5. Mark and drill the second 2" × 4".
6. Install the two carriage bolts.
7. Tighten the bolts snugly without crushing the wood fibers or drawing the head below the surface.
8. Clean the area and put away the tools.

Instructor's Score or Approval ______________________________

Job Sheet 10-6

Name ______________________________________ Date ______________

Fastening Wood with Glue and Screws

Objective

Upon completing this activity, you will have demonstrated the ability to fasten wood with glue and screws.

Tools and Materials

Job Sheet 10-6

Two pieces of wood

Wood screws

Aliphatic resin (carpenter's) glue

Putty knife

Drill

Pencil

Try square

Procedure

1. Cut the two pieces of wood so the edges match perfectly.
2. Modify the wood, if needed, to secure a complete match.
3. Drill the screw holes.
4. Apply a small bead of glue on each surface.
5. Spread the glue evenly.
6. Install the screws.
7. Remove excess glue or glue runs with the putty knife.
8. Clean the area and put away the materials.

Instructor's Score or Approval ______________________________________

Job Sheet 10-7

Name ______________________________________ Date ______________

Fastening Wood with Dowels and Glue

Objective

Upon completing this activity, you will have demonstrated the ability to fasten wood with dowels and glue.

Tools and Materials

Job Sheet 10-7

Two pieces of wood with straight edges

Two bar clamps

⅜" dowel stock

Backsaw

Sandpaper

Aliphatic resin (carpenter's) glue

Combination square

⅜" auger bit

Bit brace

Bench vise

Putty knife

Pencil

Procedure

1. Prepare perfectly fitting edges.
2. Place the two boards together face to face and then insert them vertically into the vise with the ends flush.
3. Mark across the two pieces approximately one-third the length of the board from each end.
4. Set the combination square at ⅜".
5. Make a "center mark" at ⅜" on each line made in Step 3.
6. Drill a 1⅛"-deep hole at each center mark.
7. Cut two 2" dowels.
8. Taper the ends of the dowels with sandpaper.
9. Test fit the pieces.
10. Apply a thin bead of glue on the edges to be joined and into each of the holes.
11. Assemble the pieces and then clamp them together.
12. Remove any glue runs.
13. Clean the area and put away the tools.
14. Remove the clamps after 24 hours and inspect the joint.
15. Put away the clamps.

Instructor's Score or Approval ______________________________________

UNIT 11

Finishing Wood

Finishing materials protect wood from weathering and other damage. Paints, stains, varnish, and wax are among the common wood finishes. Proper finishing requires the preparation of the wood to receive the finish. In this unit, you will learn about finishing materials and methods.

Shop Activity 11-1: Removing Dents from Wood

Class Activity 11-2: Identifying Common Sanding Materials

Shop Activity 11-3: Sanding Wood

Shop Activity 11-4: Filling Holes in Wood with Putty or Glazing Compound

Shop Activity 11-5: Filling Holes in Wood with Plastic Wood

Shop Activity 11-6: Applying a Stain to Wood

Shop Activity 11-7: Applying a Wood Sealer

Shop Activity 11-8: Applying a Finish Material to Wood

Shop Activity 11-9: Cleaning and Caring for Brushes

Job Sheet 11-1

Name ______________________________________ Date ______________

Removing Dents from Wood

Objective

Upon completing this activity, you will have demonstrated the ability to remove dents from wood.

Tools and Materials

Job Sheet 11-1

Dented wood

Soldering copper

Damp soft cloth

Procedure

1. Heat the soldering copper to 400°.
2. Place the damp cloth over the dent.
3. Apply the heat for 5 seconds until steam rises.
4. Remove the copper and cloth.
5. Examine the dent, and if still evident, repeat Steps 1 through 4.
6. Sand when the wood is dry.

Instructor's Score or Approval ______________________________________

Job Sheet 11-2

Name ________________________________ Date ____________

Identifying Common Sanding Materials

Objective

Upon completing this activity, you will have demonstrated the ability to identify the common sanding materials.

Tools and Materials

Job Sheet 11-2

Pen or pencil

Procedure

Match the materials in Column I with the descriptions of their use in Column II.

Column I	*Column II*
____ 1. Aluminum oxide	a. Used when clogging is a problem
____ 2. Emery	b. Used for hand sanding of wood
____ 3. Flint	c. Improves speed and durability
____ 4. Garnet	d. Removes rust
____ 5. Silicone carbide	e. Used for wet or dry sanding
____ 6. Steel wool	f. Used for smoothing and polishing

Instructor's Score or Approval ________________________________

Job Sheet 11-3

Name ______________________________ Date ____________

Sanding Wood

Objective

Upon completing this activity, you will have demonstrated the ability to sand wood.

Tools and Materials

Job Sheet 11-3

Medium, fine, and very fine sandpaper

Sanding block

Wood to be sanded

Bench vise

Procedure

1. Wrap the sanding block with medium sandpaper.
2. Place the wood in the bench vise so the face is above the jaws.
3. Sand with the grain, moving over the whole surface of the wood.
4. Sand each surface of the wood.
5. Change to fine sandpaper and repeat.
6. Change to very fine sandpaper and complete the sanding.
7. Inspect the surfaces and resand as needed.
8. Clean the area and put away the materials.

Instructor's Score or Approval ______________________________

Job Sheet 11-4

Name ______________________________________ Date ______________

Filling Holes in Wood with Putty or Glazing Compound

Objective

Upon completing this activity, you will have demonstrated the ability to fill holes in wood with putty or glazing compound.

Tools and Materials

Job Sheet 11-4

Wood with holes to be filled

Wood filler

Putty knife

Standard screwdriver

Tinting material

Procedure

1. Seal the wood as noted in Activity 11-7.
2. Take a small amount of putty or compound from the can.
3. Tint the material to the desired shade.
4. Using the screwdriver tip, force a bit of material into the hole.
5. Smooth and remove any excess material with the putty knife.
6. Clean the area and put away the tools and materials.
7. Let the material set overnight.
8. Apply the desired finish.

Instructor's Score or Approval ____

Job Sheet 11-5

Name ______________________________ Date ______________

Filling Holes in Wood with Plastic Wood

Objective

Upon completing this activity, you will have demonstrated the ability to fill holes in wood with plastic wood.

Tools and Materials

Job Sheet 11-5

Fine sandpaper

Wood with holes to be filled

Plastic wood of the desired color

Putty knife

Standard screwdriver

Procedure

1. Put a small amount of plastic wood on the tip of the screwdriver and close the tube or can.
2. Using the screwdriver, quickly force the material into the hole.
3. Smooth the material with the screwdriver or a putty knife.
4. Add more material if needed.
5. Recheck the can to be sure it is tightly closed.
6. Scrape off any material around the hole.
7. Wait 3 to 5 minutes for the material to dry.
8. Sand the filler and surrounding area with fine sandpaper.
9. Apply the desired finish.
10. Clean the work area and put away the materials.

Instructor's Score or Approval ______________________________

Job Sheet 11-6

Name ______________________________________ Date ______________

Applying a Stain to Wood

Objective

Upon completing this activity, you will have demonstrated the ability to apply wood stains.

Tools and Materials

Job Sheet 11-6

Wood to be stained

Stain of desired color

Clean soft rags

Brush

Procedure

1. Sand the surface to be stained according to the procedure outlined in Job Sheet 11-3.
2. Remove any sanding dust.
3. Stir the stain thoroughly.
4. Apply an even coat of stain with the brush.
5. Allow the stain to stand 5 minutes.
6. Wipe any excess stain off with the soft rags.
7. Wipe and polish lightly until no stain comes off.
8. Clean the brush and put away the materials.
9. Allow 24 hours for drying.
10. Apply the desired finish.

Instructor's Score or Approval ______________________________________

Job Sheet 11-7

Name ______________________________ Date ______________

Applying a Wood Sealer

Objective

Upon completing this activity, you will have demonstrated the ability to apply a wood sealer.

Tools and Materials

Job Sheet 11-7

Wood to be sealed

Wood sealer

Brush

Tack rag (a rag dampened with a solvent)

Cleaning solvent

Procedure

1. Sand the wood to be sealed according to the procedure outlined in Job Sheet 11-3.
2. Use a tack rag to wipe the wood.
3. To prepare the wood sealer, read and follow the directions on the container.
4. Apply an even coat with a brush.
5. Clean the brush.
6. Clean the area and put away the materials.

Instructor's Score or Approval ______________________________

Job Sheet 11-8

Name ______________________________ Date ____________

Applying a Finish Material to Wood

Objective

Upon completing this activity, you will have demonstrated the ability to apply paint or varnish to wood.

Tools and Materials

Job Sheet 11-8

Paintbrush suitable to material to be used

Finish, paint, or varnish

Wood to be finished

Tack rag

Brush-cleaning materials

Procedure

1. Sand the wood according to the procedure outlined in Job Sheet 11-3.
2. Read the product label and follow the directions to prepare the finish for use.
3. Wipe the wood with a tack rag.
4. Dip only ½" of the brush into the finish.
5. Apply the finish using back-and-forth strokes to keep the bristles together.
6. Avoid bending the bristles of the brush or spreading them apart.
7. Alternate dipping and brushing to prevent the finish from drying in the brush.
8. Coat all surfaces to be protected with an even coat.
9. Clean the brush immediately with the proper solvent. Clean and rinse further with soap and warm water. Put the materials away.

Instructor's Score or Approval ______________________________

Job Sheet 11-9

Name ______________________________ Date ____________

Cleaning and Caring for Brushes

Objective

Upon completing this activity, you will have demonstrated the ability to clean and care for brushes.

Tools and Materials

Job Sheet 11-9

Solvents

Container that's just a little bigger than the brush

Used solvent container

Paintbrush that needs cleaning

Warm water

Soap

Paper towels

Procedure

1. Remove as much material from the bristles as possible with rags or paper.
2. Pour ½" of solvent into the container.
3. Gently work the bristles up and down in the solvent.
4. Pour the solvent into the used solvent container.
5. Pour ½" of clean solvent into the container and repeat Steps 3 and 4 until the bristles are free of the finish material.
6. Wipe the excess solvent from the brush with paper towels.
7. Work the soap into a lather and wash the brush with warm water until it is clean.
8. Rinse the soap from the brush.
9. Lay the brush on the lower-left corner of a paper towel and wrap one or two turns; fold the towel down over the brush and complete the wrap. Then, twist the towel around the handle.
10. Lay the brush in a warm place to dry for approximately 24 hours.
11. Examine the brush bristles for any evidence of the finish material.

Instructor's Score or Approval ______________________________

UNIT 12

Identifying, Marking, Cutting, and Bending Metal

Most agricultural machines are made of metal, so you must learn the basics of metalworking.

Shop Activity 12-1: Identifying Metals

Shop Activity 12-2: Marking Metal

Shop Activity 12-3: Cutting Metal with a Hacksaw

Class Activity 12-4: Identifying Files

Class Activity 12-5: Identifying Metal Snips

Shop Activity 12-6: Cutting Metal with a Cold Chisel

Shop Activity 12-7: Shaping Metal with a File

Shop Activity 12-8: Bending Metal

Shop Activity 12-9: Rounding Metal

Shop Activity 12-10: Twisting Metal

Shop Activity 12-11: Bending Sheet Metal

Job Sheet 12-1

Name ______________________________ Date ____________

Identifying Metals

Objective

Upon completing this activity, you will have demonstrated the ability to identify the distinguishing characteristics of metals.

Tools and Materials

Job Sheet 12-1

Pen or pencil

Samples of metals

Procedure

Match the metals with the characteristics, have your instructor correct your answers, and then identify the numbered samples.

____	1. Aluminum	a. Reddish brown, excellent conductor
____	2. Brass	b. Soft, malleable, corrosion resistant
____	3. Bronze	c. Soft, malleable, corrosion resistant
____	4. Cast iron	d. Very malleable, corrosion resistant, silver color
____	5. Copper	e. Bright, hard, tough, corrosion resistant
____	6. Galvanized steel	f. Soft, very heavy, bluish gray
____	7. Lead	g. Silver white, good electric conductor, light, tough
____	8. Mild steel	h. Zinc-coated metal
____	9. Stainless steel	i. High carbon, heat treatable, expensive
____	10. Tin	j. Malleable, tough, rust resistant
____	11. Tool steel	k. Malleable, ductile, tough
____	12. Wrought iron	l. Forms into any shape, brittle

Sample Number

1. ____
2. ____
3. ____
4. ____
5. ____
6. ____
7. ____

8. ____
9. ____
10. ____
11. ____
12. ____
13. ____
14. ____
15. ____

Study the samples you identified incorrectly until you can identify them correctly.

Instructor's Score or Approval ________________________________

Job Sheet 12-2

Name ______________________________ Date ____________

Marking Metal

Objective

Upon completing this activity, you will have demonstrated the ability to mark metals.

Tools and Materials

Job Sheet 12-2

Soapstone

Chalk

Center punch

Hammer

Scriber

Combination square

Mild steel

Leather gloves

Safety glasses

Procedure

1. Place the combination square against a straight side of the mild steel.
2. Mark a line across the steel with the soapstone.
3. Using the scriber, mark a second line 1" from the first.
4. Put on leather gloves and safety glasses.
5. Place the tip of the center punch 1" from the edge of the metal.
6. Tap the punch lightly with the hammer.
7. Check to see if the tap mark is exactly 1" from the edge.
8. If the mark is incorrect, reset the punch.
9. With the punch in the correct spot, hit the punch with one solid blow.
10. With the chalk, make a third line on the metal.
11. Brush the metal.
12. Observe the best marks.
13. Clean the area and put away the materials.
14. Describe the performance of each marking technique:
    Soapstone: ____

    Scriber: ____

    Center punch: ____

    Chalk: ____

Instructor's Score or Approval ______________________________

Job Sheet 12-3

Name ______________________________________ Date ______________

Cutting Metal with a Hacksaw

Objective

Upon completing this activity, you will have demonstrated the ability to cut metal with a hacksaw.

Tools and Materials

Job Sheet 12-3

Mild steel

Soapstone

Combination square

Machinist's vise

Hacksaw

Procedure

1. Place the combination square on a straight edge of the mild steel.
2. Mark a line across the steel with soapstone.
3. Place the mild steel in the vise with the line ½" from the jaws of the vise.
4. Select a hacksaw blade that will have three or more teeth on the metal as it cuts.
5. Cut the metal to the mark.
6. Apply pressure only on the forward stroke.
7. Check the squareness of your cut.
8. Save the metal for Activity 12-7.
9. Clean the area and put away the materials.

Instructor's Score or Approval ____

Job Sheet 12-4

Name ______________________ Date ____________

Identifying Files

Objective

Upon completing this activity, you will have demonstrated the ability to identify files.

Tools and Materials

Job Sheet 12-4

Pen or pencil

Procedure

Correctly label the files pictured.

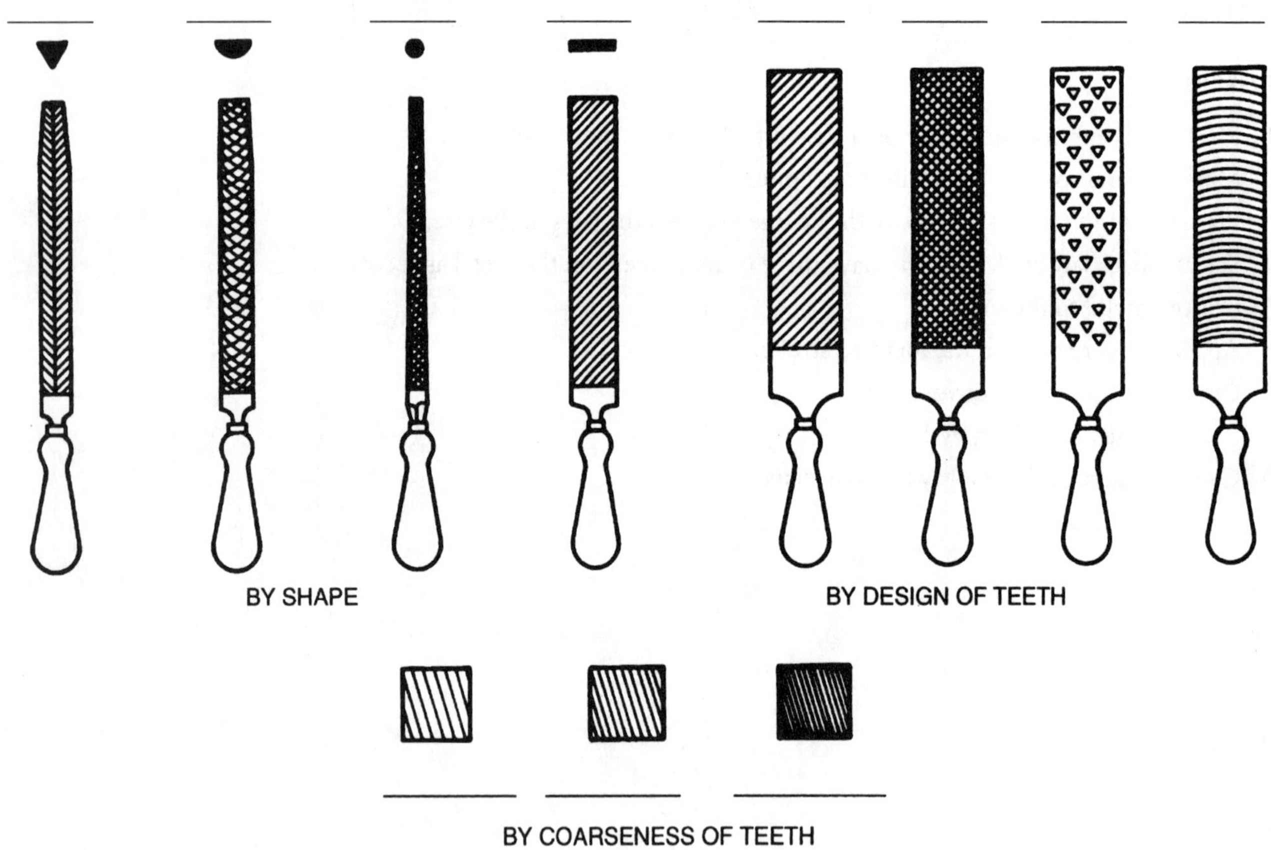

Figure 12-1 Identifying files

Instructor's Score or Approval ______________________

Job Sheet 12-5

Name ______________________________ Date __________

Identifying Metal Snips

Objective

Upon completing this activity, you will have demonstrated the ability to identify metal snips.

Tools and Materials

Job Sheet 12-5

Pen or pencil

Procedure

Correctly label the snips pictured.

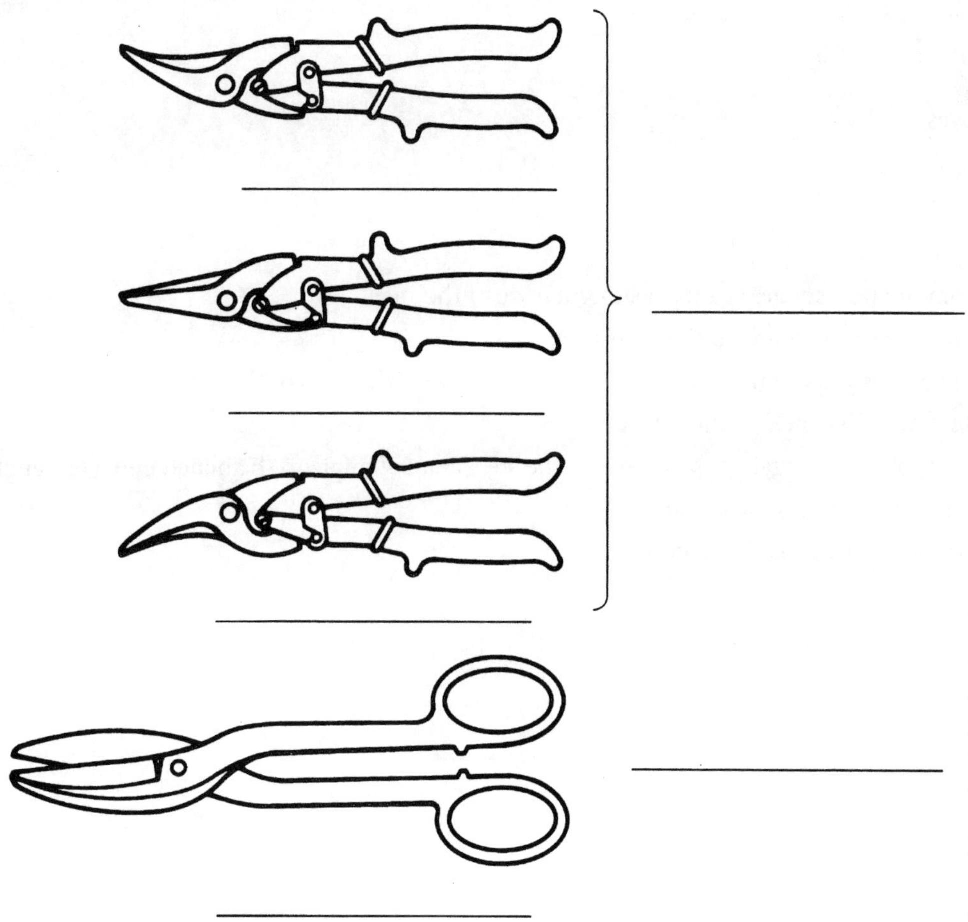

Figure 12-2 Identifying metal snips

Instructor's Score or Approval ______________________________

Job Sheet 12-6

Name ____ Date ______________

Cutting Metal with a Cold Chisel

Objective

Upon completing this activity, you will have demonstrated the ability to cut metal with a cold chisel.

Tools and Materials

Job Sheet 12-6

Mild steel

Cold chisel

Blacksmith's hammer

Machinist's vise

Combination square

Soapstone

Leather gloves

Safety glasses

Procedure

1. Place the combination square against a straight edge of the steel.
2. Mark a line across the metal with the soapstone.
3. Place the metal vertically in the vise so the mark is even with the jaws of the vise.
4. Put on the safety glasses and leather gloves.
5. Lay the heavy cold chisel against the corner of the steel at a 30° angle to the bench and a 60° angle to the metal.
6. Drive the chisel so it shears the metal.
7. Clean the area and put away the materials.

Instructor's Score or Approval ____

Job Sheet 12-7

Name ______________________________ Date ____________

Shaping Metal with a File

Objective

Upon completing this activity, you will have demonstrated the ability to shape metal with a file.

Tools and Materials

Job Sheet 12-7

Combination square

File

Metal from Activity 12-3

Machinist's vise

Scriber

Procedure

1. Check the hacksaw cut (made for Job Sheet 12-3) for squareness both crosswise and edgewise.
2. If the edge is approved for squareness, scribe a line ⅟₁₆" away from the cut.
3. If the edge is not square, scribe a line that will take it to square.
4. Hold the file tip between the thumb and index finger of one hand, and hold the handle with the other hand.
5. File to the scribe mark, applying no pressure on the backstroke.
6. Test for squareness frequently. This will tell you if you are holding the file to the end of the metal at a 90° angle.
7. Clean the area and put away the materials.

Instructor's Score or Approval ______________________________

Job Sheet 12-8

Name ____ Date _______________

Bending Metal

Objective

Upon completing this activity, you will have demonstrated the ability to bend metal.

Tools and Materials

Job Sheet 12-8

Mild steel

Combination square

Soapstone

Blacksmith's hammer

Machinist's vise

Leather gloves

Safety glasses

Procedure

1. Use the combination square to mark the metal at the desired point of the bend.
2. Place the metal vertically in the vise. The mark should be even with the top of the vise jaws.
3. Put on the leather gloves and safety glasses.
4. Use one hand to push the top of the metal in the direction of the bend while using a hammer in the other hand to strike the metal at the top of the vise.
5. Drive the metal until a 90° angle is obtained.
6. Clean the area and put away the materials.

Instructor's Score or Approval ____

Job Sheet 12-9

Name ______________________________ Date ____________

Rounding Metal

Objective

Upon completing this activity, you will have demonstrated the ability to round metal.

Tools and Materials

Job Sheet 12-9

Mild steel 1"–2" wide × ⅛" thick × 12" long

Pipe or round stock

Blacksmith's hammer

Machinist's vise

Leather gloves

Safety glasses

Procedure

1. The end of the steel to be bent should be placed against the round stock and put into the vise.
2. Put on the leather gloves and safety glasses.
3. Place one hand at the top of the steel while striking the base of the steel with the hammer. Bend the steel around the stock or pipe.
4. Release the vise and rotate the stock until the unbent metal is again in the vertical position.
5. Repeat Steps 3 and 4 until the desired roundness is attained.
6. Clean the work area and put away the materials.

Instructor's Score or Approval ______________________________

Job Sheet 12-10

Name ______________________________________ Date _______________

Twisting Metal

Objective

Upon completing this activity, you will have demonstrated the ability to twist metal.

Tools and Materials

Job Sheet 12-10

Mild steel strap ¾"–1" wide × ⅛" thick × 12+" long

12" adjustable wrench

Machinist's vise

Soapstone

Square

Procedure

1. Place the square where you want the twist to start.
2. Mark the start of the bend.
3. Measure 1½ times the width of the metal and make a second mark.
4. Place the metal in the vise so the first mark is even with the top of the vise jaws and the second mark is above the vise.
5. Slide the wrench up to the second mark.
6. Adjust the wrench to fit tightly across the width of the steel.
7. Hold the metal in one hand and twist with the other.
8. Stop at 90°.
9. Clean the area and put away the materials.

Instructor's Score or Approval ______________________________________

Job Sheet 12-11

Name ______________________________ Date ____________

Bending Sheet Metal

Objective

Upon completing this activity, you will have demonstrated the ability to bend sheet metal.

Tools and Materials

Job Sheet 12-11

Sheet metal

Angle iron

Blacksmith's hammer

Clamps

Leather gloves

Safety glasses

Procedure

1. Clamp the angle iron to the edge of the bench.
2. Place the sheet metal so the point where the bend is desired is aligned with the outside edge of the angle iron.
3. Put on the leather gloves and safety glasses.
4. Start the bend by pushing downward with your hands.
5. Continue the bend, tapping with the hammer until the bend is complete.
6. Clean the area and put away the materials.

Instructor's Score or Approval ______________________________

UNIT 13

Fastening Metal

This unit deals with three ways of fastening metals together. Metal fasteners are known as screws, bolts, and rivets. When metals are held together by melting another metal between them, the process is known as soldering or brazing. When two metals are melted together, the process is known as welding.

Shop Activity 13-1: Drilling Holes in Metal

Shop Activity 13-2: Tapping Threads in Metal

Shop Activity 13-3: Threading a Rod or Bolt

Shop Activity 13-4: Cutting and Threading Pipe

Shop Activity 13-5: Fastening Metal with Bolts and Nuts

Shop Activity 13-6: Fastening Metal with Cap Screws

Shop Activity 13-7: Fastening Metal with Rivets

Shop Activity 13-8: Fastening Metal with Sheet Metal Screws

Shop Activity 13-9: Tinning a Soldering Copper

Shop Activity 13-10: Soldering Sheet Metal

Shop Activity 13-11: Soldering Electrical Connections

Shop Activity 13-12: Soldering Copper Pipes

Job Sheet 13-1

Name ______________________________ Date __________

Drilling Holes in Metal

Objective

Upon completing this activity, you will have demonstrated the ability to drill holes in metal.

Tools and Materials

Job Sheet 13-1
Drill press
High-speed twist bit
Face shield
Leather gloves
Protective clothing
Drill press vise
Metal to be drilled
Combination square
Center punch
Ball-peen hammer
Clamp
Cutting oil
Metal file
Safety glasses

Procedure

1. Obtain permission to start this procedure.
2. Put on the face shield and safety glasses.
3. Put on the leather gloves.
4. Put on the protective clothing.
5. Measure and mark the spot to be drilled.
6. Center punch the spot with a light tap.
7. Check the position (and adjust if needed) and give the center punch one sharp blow.
8. Remove the gloves. Never wear gloves while using or adjusting the drill press.
9. Set the proper drill press speed for the size of the bit.
10. If the hole is more than ⅜" in diameter, start by drilling a ¼" pilot hole.
11. Tighten the drill in the chuck by using the chuck wrench in at least two of the holes in the chuck.
12. Remove the chuck wrench.

13. Clamp the metal in a vise or clamp it to the drill press table.
14. Lower the quill and drill the hole with even pressure.
15. Apply cutting oil to the metal to cool the bit.
16. Slacken the pressure as the drill breaks through the back of the metal.
17. Remove any burrs with the metal file.
18. Clean the area and put away the materials.

Instructor's Score or Approval ______________________________

Job Sheet 13-2

Name ______________________________ Date ______________

Tapping Threads in Metal

Objective

Upon completing this activity, you will have demonstrated the ability to thread a hole in metal.

Tools and Materials

Job Sheet 13-2

Metal to be threaded

Tap and tap wrench

Face shield

Protective clothing

Leather gloves

Cutting oil

Vise

Drill press

Safety glasses

Rags or paper towels for wiping

Procedure

1. Obtain permission to start the procedure.
2. Put on the face shield and safety glasses.
3. Put on the protective clothing.
4. Mark the location of the hole to be threaded.
5. Select the proper size drill for the tap size desired (see Figure 13-10 in the textbook).
6. Drill the pilot hole using the procedure listed on Job Sheet 13-1.
7. Select the correct tap. Use a taper tap if the hole goes through the metal. Use a plug tap and then a bottoming tap for a hole that does not go through the metal.
8. Put the tap in the tap wrench and apply some oil to the threads and the hole.
9. Hold the tap perpendicular to the hole and turn a quarter turn.
10. Reverse to break the scrap.
11. Turn one to two turns and then reverse.
12. Repeat until the threads are completed.
13. Remove the tap by turning it backward.
14. Remove the chips by wiping with a rag or paper towels.
15. Dispose of the oily rag in the proper container.
16. Clean the area and put away the materials.

Instructor's Score or Approval ______________________________

Job Sheet 13-3

Name ______________________________ Date ____________

Threading a Rod or Bolt

Objective

Upon completing this activity, you will have demonstrated the ability to thread a rod or bolt.

Tools and Materials

Job Sheet 13-3

Rod or bolt to be threaded

Die and die holder

Vise

Face shield

Cutting oil

Protective clothing

Leather gloves

File

Rags or paper towels for wiping

Safety glasses

Procedure

1. Obtain permission to start the procedure.
2. Put on the face shield and safety glasses.
3. Put on the protective clothing.
4. Put on the leather gloves.
5. Clamp the bolt or rod in the vise.
6. Slightly taper the tip of the rod with the file.
7. Select a die of the desired thread type.
8. Mount the die in the die stock.
9. Put the tapered side over the rod.
10. Apply cutting oil and turn the die a half turn.
11. Back up a quarter turn.
12. Repeat Steps 10 and 11 until a short length is threaded.
13. Remove the die and check the fit of the nut.
14. If the nut fits correctly, then continue until the desired length is threaded.
15. Remove the die and clean off any chips by wiping with a rag or paper towels.
16. Dispose of the oily rag properly.
17. Clean the area and put away the materials.

Instructor's Score or Approval ______________________________

Job Sheet 13-4

Name ______________________________ Date __________

Cutting and Threading Pipe

Objective

Upon completing this activity, you will have demonstrated the ability to cut and thread pipe.

Tools and Materials

Job Sheet 13-4

Pipe to be cut and threaded

Pipe cutter

Pipe die and stock

Pipe reamer

Face shield

Protective clothing

Leather gloves

Cutting oil

Rags for wiping

Pipe vise

Safety glasses

Procedure

1. Obtain permission to start the procedure.
2. Put on the face shield and safety glasses.
3. Put on the protective clothing.
4. Put on the leather gloves.
5. Clamp the pipe in the pipe vise.
6. Loosen the pipe cutter until the cutting wheel slides over the pipe.
7. Slide the cutter to the mark where the pipe is to be cut.
8. Turn the handle of the pipe cutter until the cutting wheel starts to bite into the pipe. ***Do not* overtighten!**
9. Rotate the cutter one or two times using cutting oil.
10. Tighten the pipe cutter a half turn and repeat Step 9.
11. Continue until the pipe is severed.
12. Install the pipe reamer and remove the ridge.
13. Taper the pipe slightly with the file to ensure that the die will fit easily.
14. Install the proper pipe die in the die stock and place the tapered edge onto the pipe end.
15. Apply moderate pressure and turn a half turn.
16. Reverse to break the chip, apply cutting oil, and then turn one to two times.

17. Repeat Steps 15 and 16 until the end of the pipe is even with the outside of the die.
18. Back the die off the pipe.
19. Clean the cutter and die by wiping with a rag.
20. Clean the area, put away the materials, and dispose of rags properly.

Instructor's Score or Approval ___________________________________

Job Sheet 13-5

Name ____ Date ______________

Fastening Metal with Bolts and Nuts

Objective

Upon completing this activity, you will have demonstrated the ability to fasten metal with bolts and nuts.

Tools and Materials

Job Sheet 13-5

Metal to be fastened

Bolts and nuts

Face shield

Protective clothing

Vise

Wrenches to fit the bolt

Lock washers

Portable drill

Appropriate-sized drill bit

Center punch

Ball-peen hammer

Combination square

Safety glasses

Procedure

1. Obtain permission to start the procedure.
2. Put on the face shield and safety glasses.
3. Put on the protective clothing.
4. Mark the metal where the bolts are to be located.
5. Following proper procedure, center punch the marks and drill the holes.
6. Countersink the holes if the bolt is a flat-head stove bolt.
7. Insert the bolts.
8. Add lock washers and nuts.
9. Tighten the nuts until the lock washers are flattened.
10. Clean the area and put away the materials.

Instructor's Score or Approval ____

Job Sheet 13-6

Name ____ Date _______________

Fastening Metal with Cap Screws

Objective

Upon completing this activity, you will have demonstrated the ability to fasten metal with cap screws.

Tools and Materials

Job Sheet 13-6

Cap screws

Lock washers

Taper, plug, and bottoming taps

Tap wrench

Portable electric drill

Twist bits of the proper sizes (See Figure 13-10 in the textbook.)

Metal to be fastened

Face shield

Protective clothing

Wrench to fit the cap screw

Vise

Center punch

Ball-peen hammer

Cutting oil

Safety glasses

Procedure

1. Obtain permission to start the procedure.
2. Put on the face shield and safety glasses.
3. Put on the protective clothing.
4. In the first piece, drill holes that are the same size as the cap screws.
5. Drill holes in the second piece with the proper size bit for the tap to be used. Drill the hole to the desired depth.
6. Tap the holes in the second piece with the taper tap using cutting oil.
7. Repeat, using the plug tap.
8. Repeat, using the bottoming tap.
9. Place lock washers on the cap screws and insert them through the first metal piece.
10. Thread the cap screws into the second piece and tighten until the lock washers are flattened.
11. Clean the area and put away the materials.

Instructor's Score or Approval ____

Job Sheet 13-7

Name ______________________________ Date ____________

Fastening Metal with Rivets

Objective

Upon completing this activity, you will have demonstrated the ability to fasten metal with rivets.

Tools and Materials

Job Sheet 13-7

Metal to be fastened

Suitable rivets

Drill and bits

Face shield

Protective clothing

Center punch

Ball-peen hammer

Vise

Cold chisel

Safety glasses

Procedure

1. Obtain permission to start the procedure.
2. Put on the face shield and safety glasses.
3. Put on the protective clothing.
4. Determine the rivet length (the thickness of the two metals plus the diameter of the rivet).
5. Cut the rivet to length with the chisel or a hacksaw.
6. Locate and center punch the location of the hole on each piece of metal.
7. Drill the proper size holes.
8. Insert the rivets and place the heads on top of the vise.
9. Peen the end until the riveting is completed.
10. Clean the area and put away the materials.

Instructor's Score or Approval ______________________________

Job Sheet 13-8

Name ______________________________ Date ____________

Fastening Metal with Sheet Metal Screws

Objective

Upon completing this activity, you will have demonstrated the ability to fasten metal with sheet metal screws.

Tools and Materials

Job Sheet 13-8

Metal to be fastened

Sheet metal screws

Face shield

Protective clothing

Punch or drill

Screwdriver

Safety glasses

Procedure

1. Obtain permission to start the procedure.
2. Put on the face shield and safety glasses.
3. Put on the protective clothing.
4. Mark the location on each piece of metal.
5. Punch or drill a hole the diameter of the screw shank into the top piece of metal.
6. Punch or drill a pilot hole into the second piece of metal.
7. Insert the screw and tighten firmly, making sure to avoid stripping the metal.
8. Clean the area and put away the materials.

Instructor's Score or Approval ______________________________

Job Sheet 13-9

Name ____ Date ______________

Tinning a Soldering Copper

Objective

Upon completing this activity, you will have demonstrated the ability to tin a soldering copper.

Tools and Materials

Job Sheet 13-9

Soldering copper

Sal-ammoniac block

Sandpaper

Wire solder

Face shield

Damp cloth

Safety glasses

Protective clothing

Procedure

1. Obtain permission to start the procedure.
2. Put on the face shield and safety glasses.
3. Put on the protective clothing.
4. Use sandpaper to remove burned materials from the soldering copper.
5. Heat the copper until it smokes when rubbed on the sal-ammoniac block.
6. Rub the copper in the sal ammoniac until it is shiny.
7. Unroll 6" of solder and touch it to the copper. If the solder doesn't run, then reheat the copper.
8. Repeat until the copper is silver colored.
9. Wipe the copper with the damp cloth; a good silver color indicates proper tinning.
10. Clean the area and put away the materials.

Instructor's Score or Approval ____

Job Sheet 13-10

Name ______________________________________ Date ______________

Soldering Sheet Metal

Objective

Upon completing this activity, you will have demonstrated the ability to solder sheet metal.

Tools and Materials

Job Sheet 13-10

Sheet metal to be soldered

Soldering copper

Wire solder

Flux

Damp cloth

Face shield

Protective clothing

Sandpaper

Safety glasses

Procedure

1. Obtain permission to start the procedure.
2. Put on the face shield and safety glasses.
3. Put on the protective clothing.
4. Use sandpaper to clean the areas to be soldered.
5. Heat the metal with the flat part of the tinned soldering copper.
6. Add flux to the metal as it heats.
7. Once the solder flows, move the copper along, following with the solder.
8. Tin both pieces of metal.
9. Place the tinned surfaces together and support them as necessary.
10. Heat the pieces directly over the joint to melt the solder.
11. Add solder as needed to fill the joint.
12. Wipe off any excess solder with the damp cloth.
13. Hold the metal together until the solder hardens.
14. Clean the area and put away the materials.

Instructor's Score or Approval ______________________________________

Job Sheet 13-11

Name ________________________________ Date ____________

Soldering Electrical Connections

Objective

Upon completing this activity, you will have demonstrated the ability to solder electrical connections.

Tools and Materials

Job Sheet 13-11
Soldering copper
Wire connections to be soldered
Vise
Pliers
Electrical tape
Rosin core solder
Face shield
Protective clothing
Safety glasses

Procedure

1. Obtain permission to start the procedure.
2. Put on the face shield and safety glasses.
3. Put on the protective clothing.
4. Twist the wires together.
5. Clamp one of the wires in a vise so the connection is located beside the vise.
6. Hold the heated copper under the bare wire and feed the solder from the top.
7. Allow the solder to cool, and then wrap the wire with electrical tape.
8. Clean the area and put away the materials.

Instructor's Score or Approval ________________________________

Job Sheet 13-12

Name ______________________________ Date ____________

Soldering Copper Pipes

Objective

Upon completing this activity, you will have demonstrated the ability to solder copper pipes.

Tools and Materials

Job Sheet 13-12

Wet cloth or rag

Copper pipes

Solder approved for copper pipe

Fine sandpaper

Flux

Gas torch

Face shield

Protective clothing

Safety glasses

Procedure

1. Obtain permission to start the procedure.
2. Put on the face shield and safety glasses.
3. Put on the protective clothing.
4. Polish the outside of the pipe and the inside of the fitting with the sandpaper.
5. Apply the paste flux to the sanded surfaces.
6. Press the pipe and fitting together.
7. Light the torch.
8. Heat the underside of the fitting with the torch.
9. Hold the solder at the joint between the pipe and the fitting.
10. Heat the fitting until the solder melts and is drawn into the joint completely around the pipe.
11. Remove the heat.
12. Cool the joint with a wet cloth or rag.
13. Pressurize the system and check for leaks.
14. Clean the area and put away the materials.

Instructor's Score or Approval ______________________________

SECTION 4

Power Tools in the Agricultural Mechanics Shop

UNIT 14

Portable Power Tools

Portable power tools are relatively inexpensive today and are in common use by homeowners as well as tradesmen and professionals. Learning how to use them safely will benefit you greatly and keep you out of harm's way. Thus, this unit emphasizes safety in their use.

Shop Activity 14-1: Safety Precautions

Shop Activity 14-2: Using a Power Drill

Shop Activity 14-3: Using a Portable Sander

Shop Activity 14-4: Using a Portable Disc Sander or Grinder

Shop Activity 14-5: Using a Finishing Sander

Shop Activity 14-6: Using a Portable Saber Saw

Shop Activity 14-7: Using a Reciprocating Saw

Shop Activity 14-8: Using a Circular Saw

Shop Activity 14-9: Using a Power Router

Job Sheet 14-1

Name ______________________________ Date ____________

Safety Precautions

Objective

Upon completing this activity, you will have demonstrated the ability to identify portable power tool safety precautions.

Tools and Materials

Job Sheet 14-1

Pen or pencil

Clipboard

Procedure

Complete the checklist for the portable power tools in the shop. Write in the name of each power tool you examine in the space provided. If you find a tool that does not meet the requirements on the checklist, notify the instructor and do not use it until the condition is corrected.

Safety/Feature/Practice	*Tool*	*Tool*	*Tool*	*Tool*
1. All power cords on metal-housed tools are three-wire safety.				
2. All power cords are free from breaks and cuts.				
3. All power cords are properly attached to tools.				
4. Blades on cutting tools are clean and sharp.				
5. Tools are clean and dry.				
6. Tools are not used in wet areas.				
7. Protective clothing is worn when tools are used.				
8. Work is supported when using portable power tools.				
9. Power tools are stored in a manner that protects the cutting edges.				
10. Permission of the instructor is required before a tool is used.				
11. Power tools are not forced or overloaded.				
12. Others are warned when tools are to be used.				
13. Vent holes in the tools are free from dirt.				
14. All switches work properly.				
15. All guards operate properly.				

Instructor's Score or Approval ______________________________

Job Sheet 14-2

Name ____ Date _______________

Using a Power Drill

Objective

Upon completing this activity, you will have demonstrated the ability to use a portable power drill.

Tools and Materials

Job Sheet 14-2

Portable power drill

Twist bit

File

Center punch

Combination square

Soapstone marker

Ball-peen hammer

Metal to be drilled

Face shield

Leather gloves

Protective clothing

Cutting oil

Safety glasses

Procedure

1. Obtain permission to start the procedure.
2. Put on the face shield and safety glasses.
3. Put on the protective clothing.
4. Put on the leather gloves.
5. Mark the drilling location in the metal.
6. Center punch the spot with a gentle tap.
7. Check the location and correct it if necessary; then give the center punch one good blow.
8. Tighten the drill bit in the chuck using at least two holes.
9. Remove the chuck wrench.
10. Hold the drill perpendicular to the metal and start the cut.
11. Use moderate pressure.
12. Frequently apply cutting oil to cool the bit.

13. Lower the pressure as the bit breaks through the underside of the metal.
14. Remove any burrs with a file.
15. Clean the area and put away the materials.

Instructor's Score or Approval ____

Job Sheet 14-3

Name ______________________________ Date ____________

Using a Portable Sander

Objective

Upon completing this activity, you will have demonstrated the ability to use a portable sander.

Tools and Materials

Job Sheet 14-3

Portable belt sander

Dust mask

Face shield

Protective clothing

Wood to be sanded

Appropriate sandpaper belts

Holding device, vise, or bench stops

Safety glasses

Procedure

1. Obtain permission to start the procedure.
2. Put on the dust mask.
3. Put on the face shield and safety glasses.
4. Put on the protective clothing.
5. Clamp the wood to be sanded.
6. Select the appropriate belt (coarse and medium) and install it.
7. Empty the dust bag if necessary.
8. Place the power cord out of the path of the sander.
9. Hold the machine above the work and turn it on. Hold the sander with both hands.
10. Place the front of the belt sander on the wood and slowly lower the rear until it rests flat on the surface.
11. Keep the sander moving back and forth over the work, sanding with the grain.
12. Raise the sander and turn it off.
13. Check the work and repeat as necessary to obtain an even, smooth surface.
14. Install a fine belt and sand to finish.
15. Empty the dust bag.
16. Clean the area and put away the materials.

Instructor's Score or Approval ______________________________

Job Sheet 14-4

Name ________________________________ Date ____________

Using a Portable Disc Sander or Grinder

Objective

Upon completing this activity, you will have demonstrated the ability to use a portable disc sander or grinder.

Tools and Materials

Job Sheet 14-4

Object to be sanded

Portable disc sander or grinder

Face shield

Dust mask

Protective clothing

Vise or clamps

Safety glasses

Procedure

1. Obtain permission to start the procedure.
2. Put on the dust mask.
3. Put on the face shield and safety glasses.
4. Put on the protective clothing.
5. Clamp the metal to be sanded.
6. Select the appropriate sanding disc or grinding wheel.
7. Place the power cord out of the path of the sander or grinder.
8. Hold the machine in both hands and turn on the switch.
9. Lower the machine to the work slowly and let the weight of the machine be the only pressure.
10. Keep the machine moving.
11. Lift the machine before turning it off, and hold it until it stops turning.
12. Check the work and repeat sanding or grinding as needed.
13. Lay the machine on the proper rest and not on the disc or grinding wheel.
14. Clean the area and put away the materials.

Instructor's Score or Approval ________________________________

Job Sheet 14-5

Name ______________________________ Date ____________

Using a Finishing Sander

Objective

Upon completing this activity, you will have demonstrated the ability to use a finishing sander.

Tools and Materials

Job Sheet 14-5

Object to be sanded

Dust mask

Face shield

Protective clothing

Finishing sander

Sandpaper

Safety glasses

Procedure

1. Obtain permission to start the procedure.
2. Put on the dust mask.
3. Put on the face shield and safety glasses.
4. Put on the protective clothing.
5. Clamp the object to be sanded.
6. Select the appropriate sandpaper and install it.
7. Start the sander and keep it moving, using coarser papers first.
8. As sanding proceeds, switch to finer sandpaper.
9. Remove dust frequently to prevent the sandpaper from clogging.
10. Clean the area and put away the materials.

Instructor's Score or Approval ______________________________

Job Sheet 14-6

Name ______________________________________ Date ______________

Using a Portable Saber Saw

Objective

Upon completing this activity, you will have demonstrated the ability to use a portable saber saw.

Tools and Materials

Job Sheet 14-6

Portable saber saw

Material to be cut

Dust mask

Face shield

Protective clothing

Vise or clamps

Safety glasses

Procedure

1. Obtain permission to start the procedure.
2. Put on the dust mask.
3. Put on the face shield and safety glasses.
4. Put on the protective clothing.
5. Clamp the work.
6. Select and install the appropriate blade for the material.
7. Set the angle required.
8. Plug in the saw and start the cut at an edge or bore a hole for an inside cut.
9. In a blind cut, put the toe of the saw's base plate on the work with the blade free. Rotate the saw toward the material until the cut is started.
10. Use firm pressure and steady movement to cut, allowing time for the blade to cut in turns and curves.
11. When finishing a cut, lessen the pressure and reduce your cutting speed.
12. Unplug the saw.
13. Remove the blade and lightly tighten the blade-holding screw.
14. Clean the area and put away the materials.

Instructor's Score or Approval ______________________________________

Job Sheet 14-7

Name ______________________________ Date ____________

Using a Reciprocating Saw

Objective

Upon completing this activity, you will have demonstrated the ability to use a reciprocating saw.

Tools and Materials

Job Sheet 14-7

Reciprocating saw

Vise or clamps

Dust mask

Face shield

Combination square

Pencil

Protective clothing

Material to be cut

Safety glasses

Procedure

1. Obtain permission to start the procedure.
2. Put on the dust mask.
3. Put on the face shield and safety glasses.
4. Put on the protective clothing.
5. Clamp the material to be cut.
6. Mark the cut.
7. Select and install the correct blade for the job.
8. Select the proper cutting speed.
9. Plug in the saw.
10. Keep the shoe against the work at all times.
11. Use the same cutting techniques as those used with the saber saw in Job Sheet 14-6.
12. Unplug the saw.
13. Remove the blade when finished.
14. Clean the area and put away the materials.

Instructor's Score or Approval ______________________________

Job Sheet 14-8

Name ______________________________ Date ____________

Using a Circular Saw

Objective

Upon completing this activity, you will have demonstrated the ability to use a circular saw.

Tools and Materials

Job Sheet 14-8

Portable circular saw

Dust mask

Face shield

Protective clothing

Wood to be cut

Vise or clamps

Safety glasses

Procedure

1. Obtain permission to start the procedure.
2. Put on the dust mask.
3. Put on the face shield and safety glasses.
4. Put on the protective clothing.
5. Clamp the material to be cut.
6. Check the blade for the job.
7. Adjust the desired angle and the depth of cut.
8. Plug in the saw.
9. If ripping, set the rip guide for the desired width so the blade cuts in the waste.
10. If crosscutting, mark the cut and then saw on the waste side of the mark.
11. If a pocket cut is to be made, use the front of the base with the blade free and lower the blade slowly into the work.
12. Hold the saw until the blade stops turning before setting it down.
13. Unplug the saw.
14. Clean the area and put away the materials.

Instructor's Score or Approval ______________________________

Job Sheet 14-9

Name ______________________________ Date ____________

Using a Power Router

Objective

Upon completing this activity, you will have demonstrated the ability to use a power router.

Tools and Materials

Job Sheet 14-9

Portable router

Appropriate router bit

Appropriate jig or guide

Extension cord

Dust mask

Face shield

Protective clothing

Scrap of test material

Material to be shaped

Vise or clamps

Safety glasses

Procedure

1. Obtain permission to start the procedure.
2. Put on the protective clothing.
3. Put on the dust mask.
4. Put on the face shield and safety glasses.
5. Select and install the proper router bit.
6. Adjust the depth of cut.
7. Clamp the test scrap securely.
8. Plug in the router.
9. Place the base plate of the router on the test material near the starting point of the cut.
10. Turn on the router and make the test cut, moving the router slowly enough to allow it to maintain speed without glazing the material.
11. Turn off the router and place it so it cannot roll.
12. Check the cut for proper shape and depth of cut.
13. Readjust the router and repeat the test cut as necessary.
14. Clamp the material to be shaped securely.
15. Turn on the router and make the cut following the guide closely.

16. Finish the cut and turn off the router.
17. Unplug the router.
18. Remove the bit.
19. Clean and store the router and the bit.
20. Clean the area and have your instructor inspect your work.

Instructor's Score or Approval ______________________________

UNIT 15 Woodworking with Power Machines

Stationary power machines are very important in modern construction. They have certain characteristics that make them more valuable than the portable power tools discussed in the previous unit. Learning to use them properly and safely will be valuable to you.

Class Activity 15-1: Identifying Safety Precautions with Power Woodworking Machines

Shop Activity 15-2: Cutting with the Band Saw

Shop Activity 15-3: Cutting with the Jigsaw

Shop Activity 15-4: Cutting with the Table Saw

Shop Activity 15-5: Checking the Radial Arm Saw Table for Levelness

Shop Activity 15-6: Crosscutting with the Radial Arm Saw

Shop Activity 15-7: Ripping with the Radial Arm Saw

Shop Activity 15-8: Using a Cutoff Saw

Shop Activity 15-9: Using the Jointer to Smooth an Edge

Shop Activity 15-10: Planing a Board to Thickness

Shop Activity 15-11: Smoothing a Board with a Stationary Sander

Job Sheet 15-1

Name ______________________________ Date ____________

Identifying Safety Precautions with Power Woodworking Machines

Objective

Upon completing this activity, you will have demonstrated the ability to identify safety precautions with power woodworking machines.

Tools and Materials

Job Sheet 15-1

Clipboard

Pen or pencil

Procedure

Complete the safety checklist by matching the items in the two columns.

____ 1. Machine use	a. Before leaving
____ 2. Safety zone	b. Before changing blades
____ 3. Push sticks	c. When moving large pieces
____ 4. Unplug the machine	d. Brush or vacuum
____ 5. Used for cleaning woodworking machines	e. For cutting short pieces of wood
____ 6. Move slowly and cautiously	f. Wear in the shop at all times
____ 7. Turn off machine	g. Only the operator is allowed here
____ 8. Overworked machine	h. Good working condition
____ 9. Get help	i. Obtain permission first
____ 10. Goggles or face shield	j. When performing all operations
____ 11. Protective clothing	k. Should be on at all times
____ 12. Machine guards	l. Around machines
____ 13. Hearing protection	m. Should not be loose or baggy
____ 14. Walk	n. Overload protection is activated
____ 15. Only use machines in	o. Worn when using larger machines

Instructor's Score or Approval ______________________________

Job Sheet 15-2

Name ____ Date _______________

Cutting with the Band Saw

Objective

Upon completing this activity, you will have demonstrated the ability to cut with the band saw.

Tools and Materials

Job Sheet 15-2

Band saw

Safety glasses

Shop coat or apron

Wood to be cut

Procedure

1. Obtain the instructor's permission to start the procedure.
2. Put on safety glasses and protective clothing.
3. Check that the blade is appropriate for the job to be performed.
4. Check the blade tension and adjust if necessary.
5. Set the guide assembly to within ⅛" of the stock.
6. Set the miter gauge to the desired angle if it is used.
7. Set the rip fence to the desired measurement if it is to be used.
8. Clear the table of all material.
9. Stand aside and start the machine.
10. Check that the blade is tracking properly.
11. Plan cuts so they go all the way through the board.
12. Slowly push the board into the blade.
13. For straight cuts, move the board in a straight line.
14. For curved cuts, rotate the board slowly as the saw cuts.
15. Complete the cuts.
16. Shut off the saw.
17. Make sure the blade stops before leaving the machine.
18. Clean the area and put away the materials.

Instructor's Score or Approval ____

Job Sheet 15-3

Name ______________________________ Date ____________

Cutting with the Jigsaw

Objective

Upon completing this activity, you will have demonstrated the ability to cut wood with the jigsaw.

Tools and Materials

Job Sheet 15-3

Jigsaw

Wood

Safety glasses

Shop coat or apron

Appropriate blade for the saw

Procedure

1. Obtain permission to use the saw.
2. Put on the safety glasses.
3. Put on the protective clothing.
4. Select a blade that is suitable for the material to be cut.
5. Unplug the jigsaw.
6. Attach the blade to the lower chuck.
7. Rotate the motor pulley until the blade reaches its highest point.
8. Attach the blade to the upper chuck.
9. Loosen the tension sleeve clamp knob.
10. Lift the tension sleeve until there is moderate tension on the blade.
11. Tighten the tension sleeve clamp knob.
12. Again rotate the pulley to check that the blade travels to its highest point without buckling. Readjust if necessary.
13. Adjust the hold-down foot tension to put slight tension on the work.
14. Plug the machine back in and turn it on.
15. Keep fingers to the side and push the wood slowly.
16. Saw on the waste side of the line.
17. Rotate the stock so the saw blade follows the line.
18. Cut out through waste and don't back out of cuts.
19. To cut inside a circle, bore a hole through the waste material. Remove the blade and place it through the pilot hole. Reinsert the blade according to the correct procedure.
20. Saw inside the circle.
21. Remove the blade.
22. Clean the area and put away the materials.

Instructor's Score or Approval ______________________________

Job Sheet 15-4

Name ______________________________________ Date _______________

Cutting with the Table Saw

Objective

Upon completing this activity, you will have demonstrated the ability to cut with a table saw.

Tools and Materials

Job Sheet 15-4

Table saw

Safety glasses

Shop coat or apron

Wood to be cut

T bevel

Procedure

1. Obtain permission to use the saw.
2. Put on the safety glasses.
3. Put on the protective clothing.
4. Unplug the saw.
5. Install the correct blade for the job.
6. Adjust the blade angle with the scale or a T bevel.
7. Adjust the blade height so only the teeth extend above the stock.
8. If ripping, remove the miter gauge from the saw. Measure from the tip of the tooth nearest the rip fence to set the desired width.
9. If crosscutting, remove the rip fence or move it out of the way.
10. Set the miter gauge to the desired angle.
11. Check that the guard assembly is in place.
12. Plug in the motor.
13. Stand aside and start the machine.
14. Push the material slowly toward the blade.
15. Hold the material firmly on both sides of the blade.
16. Use a push stick to complete the cut.
17. Shut off the machine.
18. Clean the area and put away the materials.

Instructor's Score or Approval ______________________________________

Job Sheet 15-5

Name ______________________________________ Date _______________

Checking the Radial Arm Saw Table for Levelness

Objective

Upon completing this activity, you will have demonstrated the ability to check the radial arm saw table for levelness.

Tools and Materials

Job Sheet 15-5

Radial arm saw

Face shield

Protective clothing

Procedure

1. Obtain the instructor's permission to start the procedure.
2. Put on the face shield.
3. Put on the protective clothing.
4. Unplug the saw.
5. Loosen the column clamp.
6. Turn the elevating crank until the teeth just touch the table.
7. Loosen or release the trolley lock.
8. Pull the saw across the table. If the teeth don't lightly touch the table, report the condition to the instructor and do not use the saw until the condition is corrected.

Instructor's Score or Approval ____

Job Sheet 15-6

Name ______________________________________ Date ______________

Crosscutting with the Radial Arm Saw

Objective

Upon completing this activity, you will have demonstrated the ability to crosscut with the radial arm saw.

Tools and Materials

Job Sheet 15-6

Radial arm saw

Safety glasses

Shop coat or apron

Wood to be cut

Procedure

1. Obtain the instructor's permission to start the procedure.
2. Put on the safety glasses.
3. Put on the protective clothing.
4. Mark the point of the cut.
5. Set the saw so the teeth just cut through the board and touch the table top.
6. Position the board so the saw will cut on the waste side of the mark.
7. Hold the saw with one hand and turn it on.
8. Make sure your other hand is not in line with the blade.
9. Slowly pull the saw through the board.
10. Return the saw to the column.
11. Turn off the saw and hold it until it stops turning.
12. Clean the area and put away the materials.

Instructor's Score or Approval ______________________________________

Job Sheet 15-7

Name ________________________________ Date ____________

Ripping with the Radial Arm Saw

Objective

Upon completing this activity, you will have demonstrated the ability to rip with the radial arm saw.

Tools and Materials

Job Sheet 15-7

Radial arm saw

Wood to be ripped

Safety glasses

Shop coat or apron

Pencil

Combination square

Procedure

1. Obtain permission to start the procedure.
2. Put on the safety glasses.
3. Put on the shop coat or apron.
4. Elevate the saw until the teeth clear the table.
5. Release the yoke lock and turn the saw 90° (parallel to the fence).
6. Measure the desired width of the cut by measuring from the fence to an inside tooth.
7. Lock the trolley in this position.
8. Lower the blade until the saw just touches the table.
9. Rotate the nose of the blade guard so it just clears the work by ⅛".
10. Lower the anti-kickback device to the height of the work.
11. Obtain a helper for a long board.
12. Stand aside and turn on the saw.
13. Push the wood slowly through the saw.
14. For short boards, finish with a push stick.
15. Turn off the saw and wait until it stops turning.
16. Elevate the saw and return it to the crosscutting position.
17. Clean the area and put away the materials.

Instructor's Score or Approval ________________________________

Job Sheet 15-8

Name ______________________________ Date ______________

Using a Miter (Cutoff) Saw

Objective

Upon completing this activity, you will have demonstrated the ability to use a miter, or cutoff, saw.

Tools and Materials

Job Sheet 15-8

Miter (cutoff) saw

Appropriate blade for the saw

Safety glasses

Dust mask

Protective clothing

Material to be cut

Extension cord

Procedure

1. Obtain permission to use the miter saw.
2. Put on the dust mask.
3. Put on the safety glasses.
4. Put on the protective clothing.
5. Unplug the machine and install the appropriate blade.
6. Set the saw for the desired angle of cut.
7. Plug the machine back into the outlet or extension cord.
8. While holding the saw with one hand, stand to the side and start the machine.
9. Make the cut by slowly lowering the saw into and through the stock.
10. Allow the saw to slowly rise to the parked position.
11. Turn off the saw and hold it until the blade stops turning.
12. Remove the stock.
13. Clean the area.

Instructor's Score or Approval ______________________________

Job Sheet 15-9

Name ______________________________ Date ____________

Using the Jointer to Smooth an Edge

Objective

Upon completing this activity, you will have demonstrated the ability to smooth an edge of a board with the jointer.

Tools and Materials

Job Sheet 15-9

Jointer

Board to smooth

Safety glasses

Protective clothing

Procedure

1. Obtain permission to use the jointer.
2. Put on the safety glasses.
3. Put on the protective clothing.
4. Check the blade guard for proper operation.
5. Adjust the fence to the desired angle (usually 90°).
6. Adjust the depth of cut to 1/16".
7. Plane only boards longer than 12" in length.
8. Obtain a helper if the board is more than 48" long.
9. Stand aside and start the machine.
10. Start the board on the front feed table, holding it by the top edge.
11. Slowly move the board forward while pressing down with moderate pressure.
12. Move your hands back as they near the blades.
13. Move your second hand to the rear table as the end nears the blade.
14. Use less pressure as the knives approach the end of the board.
15. Use a push stick for finishing a board under 48".
16. Reverse the direction of the board with each pass. This will keep the board straight.
17. Make the last pass with the grain.
18. Clean the area and put away the materials.

Instructor's Score or Approval ______________________________

Job Sheet 15-10

Name ______________________________ Date ____________

Planing a Board to Thickness

Objective

Upon completing this activity, you will have demonstrated the ability to plane a board to uniform thickness.

Tools and Materials

Job Sheet 15-10

Safety glasses

Protective clothing

Planer

Wood to be planed

Procedure

1. Obtain permission to start the procedure.
2. Put on the safety glasses.
3. Put on the protective clothing.
4. Check the machine guard for proper operation.
5. Identify the thickest end of the board.
6. Measure the board's thickness and set the depth of cut to ⅛" less (depth may vary depending on the hardness of the wood).
7. Select a helper when the length of the board requires assistance.
8. Place the thinner end of the board on the infeed table.
9. Stand aside and turn on the machine.
10. Push the board forward until the feed rollers take over. Never stand directly behind the board as it is fed into the planer.
11. On long boards, have the helper move to the rear of the planer once the board reaches the halfway point.
12. Plane any additional boards.
13. Raise the bed and make the second pass on the opposite side of the board.
14. Make shallower passes and finish with the grain.
15. Turn off the machine and stand by until it stops.
16. Clean the area and put away the materials.

Instructor's Score or Approval ______________________________

Job Sheet 15-11

Name ______________________________ Date ____________

Smoothing a Board with a Stationary Sander

Objective

Upon completing this activity, you will have demonstrated the ability to smooth a board with a stationary sander.

Tools and Materials

Job Sheet 15-11

Board to be sanded

Safety glasses

Protective clothing

Dust mask

Stationary sander

Procedure

1. Obtain permission to use the sander.
2. Put on the dust mask.
3. Put on the safety glasses.
4. Put on the protective clothing.
5. Check the belt to be sure it is unclogged. Replace with a clean belt if necessary.
6. Turn on the dust collector.
7. Turn on the sander.
8. Hold the wood firmly with fingers away from the belt.
9. Put the wood lightly on the belt.
10. Move the wood back and forth from side to side.
11. When using the disc, use the side that rotates down.
12. Use only moderate pressure.
13. Reduce the pressure before removing the wood.
14. In freehand sanding, use light pressure and keep the work moving at all times.
15. Turn off the sander and dust collector.
16. Wait until the sander stops.
17. Clean the area and put away the materials.

Instructor's Score or Approval ______________________________

UNIT 16 Adjusting and Maintaining Power Woodworking Equipment

To protect the investment made when purchasing power woodworking equipment, it is crucial that these machines operate safely and efficiently.

This chapter details the procedures used to properly adjust the table saw, the band saw, the jointer, and the planer.

Shop Activity 16-1: Adjusting the Table Saw

Shop Activity 16-2: Adjusting the Band Saw

Shop Activity 16-3: Adjusting the Jointer

Shop Activity 16-4: Adjusting the Planer

Job Sheet 16-1

Name ______________________________ Date ____________

Adjusting the Table Saw

Objective

Upon completing this activity, you will have demonstrated how to adjust a table saw for safe operation.

Tools and Materials

Job Sheet 16-1

Table saw

Safety glasses

Level

Screwdriver

Framing square

Rubber mallet

Wrenches

Combination square

Procedure

1. Obtain the instructor's permission to start the procedure.
2. Unplug the saw and put on the safety glasses.
3. Check the table to be sure it is level. Adjust the foot on each leg as necessary.
4. Use the framing square to check that the throat plate insert is even with the table surface. Use the set screws at the corners of the insert to make necessary adjustments.
5. Loosen the bolts that secure the table wings to the central part of the table.
6. Use the framing square to make one wing level to the central table.
7. To move the wing, tap it lightly with the rubber mallet. Tighten the bolts once the wing is level.
8. Repeat Steps 6 and 7 to adjust the other wing.
9. Use the framing square to check the saw blade angle. (Be sure the square is placed *between* two teeth on the blade.)
10. If a gap exists between the blade and the square, use the adjusting wheel to attain a 90° angle.
11. Lock the adjusting wheel in place.
12. Use the framing square to check the miter gauge for square.
13. Loosen the knob on the back of the miter gauge to adjust the angle as needed.
14. Select one tooth at the back of the blade and use the combination square to measure the distance between the tooth and the miter gauge slot in the table.
15. Turn the blade so the *same tooth* used in Step 14 is now on the front of the blade. Repeat the measurement from the tooth to the slot.
16. If the measurements made in Steps 14 and 15 are not equal, you must loosen the trunnion bolts and reposition the blade.

17. Use the same blade tooth used in Steps 14 and 15; measure from the tooth to the fence.
18. Loosen the handle or knob on the fence and reposition the fence until the two measurements in Step 17 are the same. Lock the fence into place.
19. Plug in the saw and put away the tools.

Instructor's Score or Approval ______________________________

Job Sheet 16-2

Name ______________________________ Date ____________

Adjusting the Band Saw

Objective

Upon completing this activity, you will have demonstrated how to adjust a band saw for safe operation.

Tools and Materials

Job Sheet 16-2

Band saw

Safety glasses

Wrenches

Try square or combination square

Procedure

1. Obtain the instructor's permission to start the procedure.
2. Unplug the saw and put on the safety glasses.
3. Remove the upper blade guard.
4. Use the combination square to make sure that the blade is perpendicular to the table.
5. If the angle measured in Step 4 is not 90°, use the trunnion to adjust the angle of the table.
6. Remove the lower blade guard.
7. Examine the rubber tire surrounding the drive wheels. Inform your instructor of any grooves or excessive wear.
8. Check the blade position on the upper wheel. The blade should track in the center of the tire when the wheel is spun.
9. To adjust the blade tracking, make minor turns with the knob located on the back of the upper wheel.
10. After adjusting the wheel, turn it by hand. Once correct, the blade should not wander.
11. If the blade wanders, repeat Steps 9 and 10 as necessary.
12. Check the blade guides for proper clearance. The blade should not make contact with the guides when the saw is not in use. Too much clearance also is a problem. To gauge proper clearance, slide a magazine advertisement card between the blade and the guides. The card should make slight contact with the blade and the guide.
13. If necessary, adjust guide clearance by loosening the set screw(s) on each guide and repositioning it/them.
14. Check the blade tension by pushing on the blade halfway between the upper wheel and the blade guides. The blade should move about ¼".
15. Use the tension knob located behind the top guide wheel to adjust the tension as necessary.
16. Replace all guards and covers and remove all tools.
17. Plug the machine in and run the saw to check that the blade tracks correctly.

Instructor's Score or Approval ______________________________

Job Sheet 16-3

Name ______________________________ Date ____________

Adjusting the Jointer

Objective

Upon completing this activity, you will have demonstrated how to adjust the jointer for safe operation.

Tools and Materials

Job Sheet 16-3

Jointer

Safety glasses

Metal straight edge

Try square

Procedure

1. Obtain the instructor's permission to start the procedure.
2. Unplug the jointer and put on the safety glasses.
3. Place the straight edge on the outfeed table.
4. The infeed table should be 1⁄16" lower than the outfeed table.
5. Measure the gap between the straight edge and the infeed table to determine if it is correct.
6. *Carefully* rotate the cutting knives until they are at their highest point.
7. Use the straight edge to check that the outfeed table is level with the knives.
8. Adjust the outfeed table if necessary.
9. Use the try square to check the angle between the fence and the table.
10. If necessary, use the tilt lock on the rear of the fence to achieve a 90° angle.

Instructor's Score or Approval ______________________________

Job Sheet 16-4

Name ______________________________ Date ____________

Adjusting the Planer

Objective

Upon completing this activity, you will have demonstrated how to adjust the planer for safe operation.

Tools and Materials

Job Sheet 16-4

Planer with operator's manual

Safety glasses

Wood block approximately 12" long × 4" wide × 2" thick

Metal straight edge

Ruler/tape measure

Procedure

1. Obtain the instructor's permission to start the procedure.
2. Unplug the planer and put on the safety glasses.
3. Lower the planer table until you can insert the wood block.
4. Raise the table until the cutter head comes into contact with the board.
5. Slide the board from left to right, checking that it has the same clearance on both sides.
6. If necessary, level the cutter head and/or table according to the manufacturer's instructions.
7. Lower the table to its lowest position and remove the board.
8. Place the straight edge through the planer so that it rests on both feed rollers.
9. Adjust both rollers so that they are level with the planer table.
10. Remove the dust cover from the cutter head.
11. Measure the distance between the tip of the cutter and the chip breaker. This distance should be 1⁄16" if operating with a dust collector, or ¼" if operating without a dust collector.
12. Adjust the chip breaker if necessary.
13. Reassemble the planer.
14. Plug the planer in.
15. Run a board through the planer.
16. If the planed surface appears to have a ripple effect, then the cutter head needs to have the knives reset by a professional.
17. Clean the area and put away the tools.

Instructor's Score or Approval ______________________________

UNIT 17

Metalworking with Power Machines

The power machines covered in this unit are the drill press, grinder, power hacksaw, metal cutting shear, and metal break.

Shop Activity 17-1: Boring Holes with the Drill Press

Shop Activity 17-2: Shaping Metal with the Grinding Wheel

Shop Activity 17-3: Cleaning Metal with the Wire Wheel

Shop Activity 17-4: Dressing the Grinding Wheel

Shop Activity 17-5: Cutting Stock to Length with the Power Hacksaw

Shop Activity 17-6: Cutting Mild Steel with the Power Shear

Job Sheet 17-1

Name ______________________________ Date ____________

Boring Holes with the Drill Press

Objective

Upon completing this activity, you will have demonstrated the ability to bore holes with the drill press.

Tools and Materials

Job Sheet 17-1
Face shield
Protective clothing
Leather gloves
Drill bit
Metal to be drilled
Center punch
Ball-peen hammer
Combination square
Cutting oil
Chuck wrench
Metal file
Drill press vise
Safety glasses

Procedure

1. Obtain permission to use the drill press.
2. Put on the face shield and safety glasses.
3. Put on the protective clothing.
4. Put on the leather gloves if the metal is sharp or irregular in shape. *Never* reach near the drill bit when wearing leather gloves.
5. Use the combination square to mark the spot to be drilled.
6. Place the center punch on the mark to be drilled and give it a light tap.
7. Measure the mark for accuracy.
8. Correct if needed and give the center punch one sharp blow.
9. Place the metal in the drill press vise or clamp securely.
10. Use a pilot drill if the finished hole will be more than ⅜" in diameter.
11. Install the drill and tighten by using at least two holes in the chuck.
12. Remove the chuck wrench.
13. Turn on the machine.
14. Lower the feed handle and start the hole using steady pressure.

15. Regularly add cutting oil to keep the bit cool.
16. Reduce pressure as the bit breaks through the underside of the metal.
17. Raise the drill.
18. Turn off the machine.
19. Remove any burrs with a file.
20. Clean the area and put away the materials.

Instructor's Score or Approval ______________________________

Job Sheet 17-2

Name ______________________________ Date ____________

Shaping Metal with the Grinding Wheel

Objective

Upon completing this activity, you will have demonstrated the ability to shape metal with a grinding wheel.

Tools and Materials

Job Sheet 17-2

Face shield

Protective clothing

Metal to be shaped

Container of water

Grinding wheel

Safety glasses

Procedure

1. Obtain permission to use the grinding wheel.
2. Put on the face shield and safety glasses.
3. Put on the protective clothing.
4. Check the wheel for nicks or chips.
5. Only use grinding wheels free of nicks or chips.
6. Adjust the tool rest to about 1/16" clearance.
7. Set the tool rest so that its surface is even with the center of the motor shaft.
8. If bevel grinding, set the appropriate angle.
9. Grinders create sparks, so be sure the area is clean of any type of flammable materials.
10. Stand to the side and turn on the grinder.
11. Hold the metal firmly on the tool rest. (If the metal is relatively small, use a pair of pliers to hold it.)
12. Move the metal back and forth in the needed shaping pattern.
13. Cool the metal as needed with water.
14. Turn off the grinding wheel.
15. Wait until the grinding wheel stops running.
16. Clean the area and put away the material.

Instructor's Score or Approval ______________________________

Job Sheet 17-3

Name ______________________________ Date ____________

Cleaning Metal with the Wire Wheel

Objective

Upon completing this activity, you will have demonstrated the ability to clean metal with a wire wheel.

Tools and Materials

Job Sheet 17-3

Face shield

Protective clothing

Metal to be cleaned

Wire wheel

Dust mask

Safety glasses

Procedure

1. Obtain permission to use the wire wheel.
2. Put on the face shield and safety glasses.
3. Put on the protective clothing.
4. Put on the dust mask.
5. Stand aside and turn on the machine.
6. Hold the metal firmly with both hands. (If the metal piece is small, use pliers to hold it.)
7. Move the metal back and forth using moderate pressure until it is clean.
8. Turn off the machine.
9. Wait until the machine stops running.
10. Clean the area and put away the materials.

Instructor's Score or Approval ______________________________

Job Sheet 17-4

Name ____ Date _______________

Dressing the Grinding Wheel

Objective

Upon completing this activity, you will have demonstrated the ability to dress the grinding wheel.

Tools and Materials

Job Sheet 17-4

Face shield

Protective clothing

Leather gloves

Grinding wheel

Dust mask

Wheel dresser

Combination square

Safety glasses

Procedure

1. Obtain permission to use the grinding wheel.
2. Put on the face shield and safety glasses.
3. Put on the protective clothing.
4. Put on the leather gloves.
5. Put on the dust mask.
6. Stand aside and turn on the machine.
7. Hold the wheel dresser square with the face of the grinding wheel.
8. Move the dresser forward until its wheels make contact with the grinding wheel.
9. Move the dresser back and forth slowly for 30 seconds to clean and true the wheel.
10. Remove the dresser and turn off the grinder.
11. Readjust the tool rest to $\frac{1}{16}$" clearance.
12. Rotate the wheel by hand to check the roundness.
13. Use the combination square to check the squareness of the face.
14. Repeat dressing if needed.
15. Clean the area and put away the materials.

Instructor's Score or Approval ____

Job Sheet 17-5

Name ____ Date _______________

Cutting Stock to Length with the Power Hacksaw

Objective

Upon completing this activity, you will have demonstrated the ability to cut stock to length with the power hacksaw.

Tools and Materials

Job Sheet 17-5

Protective clothing

Leather gloves

Stock to be cut

Power hacksaw

Combination square

Safety glasses

Procedure

1. Obtain permission to use the saw.
2. Put on the safety glasses.
3. Put on the protective clothing.
4. Put on the leather gloves.
5. Use the square to mark the metal where it will be cut.
6. Check and be certain that the type of blade on the machine matches the cut to be made.
7. Raise the saw blade.
8. Locate the material in the vise so the blade cuts in the waste. Tighten the vise.
9. Turn on the machine and lower the blade slowly.
10. Turn on the coolant pump.
11. Stand to the side while the cut is made.
12. Turn off the machine when the cut is finished.
13. Remove the metal and the scrap.
14. Clean the area and put away the materials.

Instructor's Score or Approval ____

Job Sheet 17-6

Name ______________________________ Date ____________

Cutting Mild Steel with the Power Shear

Objective

Upon completing this activity, you will have demonstrated the ability to cut mild steel with the power shear.

Tools and Materials

Job Sheet 17-6

Protective clothing

Leather gloves

Power shear

Combination square

Metal to be cut

Safety glasses

Procedure

1. Obtain permission to use the power shear.
2. Put on the safety glasses.
3. Put on the protective clothing.
4. Put on the leather gloves.
5. Use the combination square to mark the metal.
6. Raise the shear handle.
7. Align the metal so the shear will cut in the waste.
8. Support the metal so it is level.
9. Lower the shear handle slowly and safely.
10. Remove the metal and the scrap.
11. Clean the area and put away the materials.
12. Secure the shear handle so it cannot be operated accidentally.

Instructor's Score or Approval ______________________________

SECTION 5

Project Planning

Unit 18

Sketching and Drawing Projects

Unit 19

Figuring a Bill of Materials

Unit 20

Selecting, Planning, and Building a Project

UNIT 18 Sketching and Drawing Projects

The time one spends in planning is amply rewarded in the construction phase. Learning how to sketch and draw will help you plan the details of a project.

A sketch or a drawing is a representation of the project that shows the visible features of the object so you can tell how it should look. These also help you determine the amounts and kinds of materials needed to make the object.

Class Activity 18-1: Identifying the Commonly Used Drawing Materials

Class Activity 18-2: Identifying the Commonly Used Lines in Sketching or Drawing

Class Activity 18-3: Understanding Sketches and Drawings

Class Activity 18-4: Practicing the Use of a Scale

Class Activity 18-5: Making a Three-View Drawing

Job Sheet 18-1

Name ______________________________ Date ____________

Identifying the Commonly Used Drawing Materials

Objective

Upon completing this activity, you will have demonstrated the ability to identify the commonly used drawing materials.

Tools and Materials

Job Sheet 18-1

Pen or pencil

Procedure

Complete the checklist by matching the drawing object with its use.

____	1. Measure or draw angles	a. T-square
____	2. Draw curved lines	b. Triangle
____	3. Draw parallel lines	c. Protractor
____	4. Makes the lines	d. Eraser
____	5. Correct mistakes	e. Pencil
____	6. Proportional drawing	f. Drawing board
____	7. Drawing surface	g. Scale
____	8. 90° or angled lines	h. Ruler
____	9. Measure lines	i. Compass

Instructor's Score or Approval ______________________________

Job Sheet 18-2

Name ______________________________ Date ____________

Identifying the Commonly Used Lines in Sketching or Drawing

Objective

Upon completing this activity, you will have demonstrated the ability to identify the commonly used lines in sketching or drawing.

Tools and Materials

Job Sheet 18-2

Pencil

Procedure

1. Complete the activity by matching the line titles to the appropriate descriptions.

____	1. Border line	a. A solid line with an arrow that explains a specific feature of an object
____	2. Object line	b. A series of dashes indicating a hidden edge
____	3. Hidden line	c. A long–short–long line indicating the center of a round object
____	4. Dimension line	d. A heavy solid line parallel to the edges of the paper
____	5. Extension line	e. A solid zigzag line used to show that the illustration stops but the object does not stop
____	6. Break line	f. A solid line showing visible edges of an object
____	7. Centerline	g. A solid line showing the exact area specified by a dimension
____	8. Leader line	h. A solid line with arrowheads at the ends to indicate a measurement

2. In the space to the right, draw the following lines:

   A border line

   An object line

   A hidden line

   A dimension line

   An extension line

   A center line

   A break line

   A leader line

Instructor's Score or Approval ______________________________

Job Sheet 18-3

Name ______________________________ Date ____________

Understanding Sketches and Drawings

Objective

Upon completing this activity, you will have demonstrated an understanding of sketches and drawings.

Tools and Materials

Job Sheet 18-3

Pen or pencil

Procedure

Answer the following questions:

1. How do sketches and drawings differ?

____

____

____

2. What is a scale?

____

____

3. Why is a scale a valuable drawing aid?

____

____

4. What is a scale drawing?

____

____

5. Describe a pictorial drawing.

____

____

6. What is a three-view drawing?

____

____

____

7. Why is it necessary to draw three views of a project?

____

____

____

Instructor's Score or Approval ______________________________

Job Sheet 18-4

Name ______________________________ Date ____________

Practicing the Use of a Scale

Objective

Upon completing this activity, you will have demonstrated the ability to use a scale.

Tools and Materials

Job Sheet 18-4

Pencil

Scale

Procedure

In the space provided, draw the desired lines to the scale indicated.

Scale Line

⅛" = 1' 3'

¼" = 1' 5'

3/16" = 1' 8'

Instructor's Score or Approval ______________________________

Job Sheet 18-5

Name ______________________________________ Date _____________

Making a Three-View Drawing

Objective

Upon completing this activity, you will have demonstrated the ability to make a three-view drawing.

Tools and Materials

Job Sheet 18-5

Pencil

Eraser

Scale

T-square

Triangle

Drafting paper

Masking tape

Drawing board

Procedure

Because this is your first drawing experience, you will reproduce the three-view drawing in Figure 18-19 in the textbook with a different scale, allowing you to follow the same procedure.

1. Place the drafting paper on the drawing board using the T-square to align it with the board.
2. Tape the corners of the paper to the drawing board.
3. Draw a heavy line 1" from each edge of the paper for the border lines.
4. Measure ½" up from the bottom border line and draw a heavy line for the title block.
5. Use the T-square and triangle to divide the title block at 2½" and 4½" from the left end, and 2½" from the right end.
6. Label the drawing with your name, the date, the title: Show box, and the scale 1½" = 1' in the title blocks.
7. Make a light mark 2" from the left border line and draw a very light vertical line from the top border to the bottom border.
8. Measure to the right 6" and repeat.
9. Measure 2" and 5" to the right and draw very light lines that only extend 5" up from the title block.
10. Measure 2" up from the title block. Draw a light line from the first vertical line to the second, and from the third to the fourth, parallel to the title block.
11. Measure 4¼" up from the title block and repeat the lines in Step 10.
12. Measure 6¼" up from the title block and draw a light line between the first two vertical lines.
13. Measure 9¼" up from the title block and draw a light line between the first two vertical lines. This completes the object lines for the front, end, and top views. Now, you will add the details to each view.
14. Measure ⅛" in from the left and right sides of the **front view**, and draw a broken vertical line through the **front** and **top views**.

15. Measure ⅛" up from the bottom of the **front view**, and draw a broken horizontal line through the **front view** and the broken lines drawn previously.
16. Continue this broken line through the **end view**.
17. Measure down ⅛" from the top of the **front view**, and draw a solid horizontal line from solid line to solid line to represent the lid. Erase any broken lines extending into this space.
18. Repeat Step 17 for the **end view**.
19. Measure ⅛" up from the bottom of the **top view**, and draw a broken line between the two broken lines drawn in Step 14.
20. Measure down from the top of the **top view** and repeat Step 19.
21. Measure 1⁄16" from the bottom of the **top view**, and draw a broken line from the left and right object lines to represent the ⅜" rabbet.
22. Erase any broken lines drawn in Step 14 that extend beyond the line drawn in Step 21.
23. Measure 1⁄16" down from the top of the **top view** and repeat Steps 21 and 22.
24. Measure ⅛" in from the left and right sides of the **end view** and draw a vertical broken line between the bottom and the lid.
25. Erase any broken lines extending beyond these two broken lines.
26. Measure 1⁄16" in from the left and right sides of the **end view**, and draw a solid line from the bottom to the lid to represent the visible end of the rabbet joint.
27. Measure 1½" and 17⁄16" up from the bottom at each end of the **front view**, and mark where the handles are to be attached.
28. Now, draw a ⅛" horizontal line that extends **out** from the box at each of the marks made in Step 27.
29. Draw vertical lines to connect the outside ends of the lines drawn in Step 28. The handles are now complete on the **front view**.
30. Locate the vertical center of each end of the **top view**. Measure ⅜" up and down from each of those locations and draw ⅛" horizontal lines outward from each side of the top.
31. With the T-square and triangle, complete the vertical lines between the lines in Step 30.
32. Locate the horizontal center of the **end view** and make a mark ⅜" to each side of the center.
33. Extend the horizontal handle lines from the **front view**, between the two marks made in Step 32.
34. Complete the end view of the handles by drawing vertical lines between the lines drawn in Step 33.
35. Clean the area and put away the materials.

Instructor's Score or Approval ______________________________

UNIT 19

Figuring a Bill of Materials

Before you can build a project, you must obtain the materials needed for construction. Knowing how to figure a bill of materials is an essential skill in project construction. In this unit, you learn how to use the drawings created in the last unit to figure a bill of materials.

Class Activity 19-1: Identifying the Components of a Bill of Materials

Class Activity 19-2: Figuring the Bill of Materials for the Show Box

Class Activity 19-3: Calculating Board Feet

Job Sheet 19-1

Name ______________________________ Date ____________

Identifying the Components of a Bill of Materials

Objective

Upon completing this activity, you will have demonstrated the ability to identify the components of a bill of materials.

Tools and Materials

Job Sheet 19-1

Pen or pencil

Procedure

Answer the following questions:

1. What is a bill of materials?

___

___

2. What are the items of information contained in the bill?

___

___

3. What is a board foot of lumber?

___

4. What are the two methods used to calculate board feet?

___

___

___

5. What lumber product(s) is/are sold by the sheet or square foot?

___

___

Instructor's Score or Approval ______________________________

Job Sheet 19-2

Name ______________________________ Date ____________

Figuring the Bill of Materials for the Show Box

Objective

Upon completing this activity, you will have demonstrated the ability to figure a simple bill of materials.

Tools and Materials

Job Sheet 19-2

Pen or pencil

Drawing in Figure 18-19 in the textbook

Procedure

Complete the following materials list form.

Item	*Quantity*	*Size*	*Bd. Ft.*	*Cost**
Top	____	____	____	$____
Bottom	____	____	____	____
Sides	____	____	____	____
Ends	____	____	____	____
Handles	____	____	____	____
Hinges	____	____	(Not shown)	____
Hasp	____	____	(Not shown, optional)	____
Finish	____	____		____
	Totals**		____	____

* Obtain the prices from your instructor.

** Nails, screws, and glue are not included.

Instructor's Score or Approval ______________________________

Job Sheet 19-3

Name ______________________________ Date ____________

Calculating Board Feet

Objective

Upon completing this activity, you will have demonstrated the ability to calculate board feet.

Tools and Materials

Job Sheet 19-3

Pen or pencil

Procedure

Calculate the number of board feet in the following list of lumber materials for a hayrack.

Item	*Number*	*Size*	*Bd. Ft. Each*	*Total Bd. Ft.*
Sills	2	2" × 8" × 14'	___	___
Joists	8	2" × 6" × 8'	___	___
Side Rails	2	2" × 4" × 14'	___	___
Floor	16	1" × 6" × 14'	___	___
Standards	4	2" × 4" × 6'	___	___
Standards	4	2" × 4" × 5'	___	___
Ends	10	1" × 6" × 8'	___	___
Braces	4	1" × 6" × 6'	___	___
		Total Board Feet	___	___

Instructor's Score or Approval ______________________________

UNIT 20

Selecting, Planning, and Building a Project

A good project will help you develop and measure your skills learned in agricultural mechanics. You may decide to use a plan, create one of your own, or modify an existing plan. Your decisions on projects should be discussed with your instructor.

Class Activity 20-1: Agricultural Mechanics Competency Inventory

Class Activity 20-2: Identifying Project Selection Criteria

Class Activity 20-3: Selecting a Project

Class Activity 20-4: Preparing a Plan

Class Activity 20-5: Identifying Design Principles

Job Sheet 20-1

Name ______________________________ Date ____________

Agricultural Mechanics Competency Inventory

Objective

Upon completing this activity, you will have inventoried your agricultural mechanics competencies.

Tools and Materials

Job Sheet 20-1

Pen or pencil

Competency checklist

Procedure

Complete the competency checklist on the next page.

AGRICULTURAL MECHANICS COMPETENCY INVENTORY

INSTRUCTION: Check the column that best describes your personal level of experience in each area of competency.

	Level of Experience			
	1	*2*	*3*	*4*
	Have Done	*Am Familiar with*	*Will Learn This Year*	*No Skill; Will Not Learn This Year*
1. Practice Safety and Shop Organization				
a. Work safely with hand tools	❑	❑	❑	❑
b. Work safely with power tools	❑	❑	❑	❑
c. Follow safety rules in shop	❑	❑	❑	❑
d. Maintain safe work areas	❑	❑	❑	❑
e. Eliminate fire hazards	❑	❑	❑	❑
f. Select and use fire extinguishers	❑	❑	❑	❑
g. Interpret product labels	❑	❑	❑	❑
h. Act correctly in an emergency	❑	❑	❑	❑
i. Store materials correctly	❑	❑	❑	❑
j. Clean the shop effectively	❑	❑	❑	❑
2. Identify and Fit Agricultural Tools				
a. Identify tools and equipment	❑	❑	❑	❑
b. Sharpen hand tools	❑	❑	❑	❑
c. Sharpen twist drills	❑	❑	❑	❑
d. Fit handles for hammers and axes	❑	❑	❑	❑
e. Fit handles for hoes and shovels	❑	❑	❑	❑
f. Dress grinding wheels	❑	❑	❑	❑

	Level of Experience			
	1	*2*	*3*	*4*
	Have Done	*Am Familiar with*	*Will Learn This Year*	*No Skill; Will Not Learn This Year*
3. Maintain and Service a Home Shop				
a. Clean and maintain an orderly shop	❑	❑	❑	❑
b. Observe safety regulations	❑	❑	❑	❑
c. Place and use fire-fighting equipment	❑	❑	❑	❑
d. Inventory tools and equipment	❑	❑	❑	❑
e. Plan shop layout	❑	❑	❑	❑
f. Select shop site	❑	❑	❑	❑
4. Apply Farm Carpentry Skills				
a. Select lumber for a job	❑	❑	❑	❑
b. Operate power saws	❑	❑	❑	❑
c. Operate a jointer	❑	❑	❑	❑
d. Operate a planer	❑	❑	❑	❑
e. Saw lumber to dimension	❑	❑	❑	❑
f. Bore holes	❑	❑	❑	❑
g. Glue wood	❑	❑	❑	❑
h. Drive and remove nails	❑	❑	❑	❑
i. Set screws and install bolts	❑	❑	❑	❑
5. Properly Use Paint and Paint Equipment				
a. Select paints or preservatives	❑	❑	❑	❑
b. Compute area for painting	❑	❑	❑	❑
c. Prepare wood surface	❑	❑	❑	❑
d. Prepare metal surfaces	❑	❑	❑	❑
e. Apply paint with a brush or roller	❑	❑	❑	❑
f. Clean and store paintbrushes	❑	❑	❑	❑
g. Mask areas prior to painting	❑	❑	❑	❑
h. Apply paint with a spray can	❑	❑	❑	❑
i. Clean and store spray gun	❑	❑	❑	❑
6. Operate an Arc Welder				
a. Practice safety in arc welding	❑	❑	❑	❑
b. Prepare metal for welding	❑	❑	❑	❑
c. Determine welder settings	❑	❑	❑	❑
d. Select electrodes	❑	❑	❑	❑
e. Operate AC and DC welders	❑	❑	❑	❑
f. Strike an arc	❑	❑	❑	❑
g. Run flat beads	❑	❑	❑	❑
h. Make weld joints	❑	❑	❑	❑
i. Weld horizontally	❑	❑	❑	❑
j. Weld vertically	❑	❑	❑	❑
k. Weld overhead	❑	❑	❑	❑

	Level of Experience			
	1 *Have Done*	2 *Am Familiar with*	3 *Will Learn This Year*	4 *No Skill; Will Not Learn This Year*
l. Cut with arc	❑	❑	❑	❑
m. Weld cast iron	❑	❑	❑	❑
n. Hard surface steel	❑	❑	❑	❑
7. Operate an Oxyacetylene Welder				
a. Check for gas leaks	❑	❑	❑	❑
b. Turn equipment on and off	❑	❑	❑	❑
c. Adjust equipment	❑	❑	❑	❑
d. Change cylinders	❑	❑	❑	❑
e. Cut with cutting torch	❑	❑	❑	❑
f. Choose proper tip sizes	❑	❑	❑	❑
g. Braze thin metal	❑	❑	❑	❑
h. Run beads	❑	❑	❑	❑
i. Make butt welds	❑	❑	❑	❑
8. Perform Skills in Hot and Cold Metal Work				
a. Identify metals	❑	❑	❑	❑
b. Cut with hand hacksaw	❑	❑	❑	❑
c. Cut with cold chisel	❑	❑	❑	❑
d. Bend metal	❑	❑	❑	❑
e. Cut with tin snips	❑	❑	❑	❑
f. Drill holes	❑	❑	❑	❑
g. Cut threads	❑	❑	❑	❑
h. Use files	❑	❑	❑	❑
i. Solder metal	❑	❑	❑	❑
j. Use power hacksaw	❑	❑	❑	❑
k. Heat metal with torch	❑	❑	❑	❑
l. Operate gas forge	❑	❑	❑	❑
9. Perform Skills in Concrete and Masonry Work				
a. Build and prepare forms	❑	❑	❑	❑
b. Treat forms	❑	❑	❑	❑
c. Reinforce concrete	❑	❑	❑	❑
d. Test aggregates for impurities	❑	❑	❑	❑
e. Mix concrete	❑	❑	❑	❑
f. Pour concrete	❑	❑	❑	❑
g. Embed bolts in concrete	❑	❑	❑	❑
h. Finish concrete	❑	❑	❑	❑
i. Protect concrete while curing	❑	❑	❑	❑
j. Trowel concrete	❑	❑	❑	❑
k. Remove forms	❑	❑	❑	❑

	Level of Experience			
	1	*2*	*3*	*4*
	Have Done	*Am Familiar with*	*Will Learn This Year*	*No Skill; Will Not Learn This Year*
l. Lay concrete blocks	❑	❑	❑	❑
m. Drill holes in concrete	❑	❑	❑	❑
10. Operate and Maintain Farm Engines				
a. Read and follow operator's manual	❑	❑	❑	❑
b. Start and operate engines	❑	❑	❑	❑
c. Change oil and oil filters	❑	❑	❑	❑
d. Service air and fuel filters	❑	❑	❑	❑
e. Maintain battery water	❑	❑	❑	❑
f. Maintain and operate small gas engines	❑	❑	❑	❑
g. Identify engine components and systems	❑	❑	❑	❑
h. Remove and connect battery cables	❑	❑	❑	❑
i. Charge batteries	❑	❑	❑	❑
j. Read battery hydrometer	❑	❑	❑	❑
k. Test batteries	❑	❑	❑	❑
l. Troubleshoot fuel problems	❑	❑	❑	❑
m. Troubleshoot ignition problems	❑	❑	❑	❑
11. Perform Minor Tune-up and Repair of Farm Engines				
a. Clean, gap, and replace spark plugs	❑	❑	❑	❑
b. Adjust carburetor mixture and speed crews	❑	❑	❑	❑
c. Install gap breaker points	❑	❑	❑	❑
d. Set dwell with meter	❑	❑	❑	❑
e. Install condensers	❑	❑	❑	❑
f. Time engines using timing light	❑	❑	❑	❑
g. Disassemble and reassemble distributors	❑	❑	❑	❑
h. Measure compression	❑	❑	❑	❑
i. Adjust valves	❑	❑	❑	❑
j. Set float levels in carburetors	❑	❑	❑	❑
k. Rebuild carburetors	❑	❑	❑	❑
l. Operate engine analyzers	❑	❑	❑	❑
m. Test and replace coils	❑	❑	❑	❑
n. Clean and install brushes	❑	❑	❑	❑
o. Clean commutator bars	❑	❑	❑	❑
12. Perform Electrification Skills				
a. Demonstrate safe practices in electrical wiring	❑	❑	❑	❑
b. Select correct fuse sizes	❑	❑	❑	❑
c. Replace fuses	❑	❑	❑	❑
d. Make splices	❑	❑	❑	❑
e. Repair electrical cords	❑	❑	❑	❑
f. Wire single-pale switches	❑	❑	❑	❑

	Level of Experience			
	1	*2*	*3*	*4*
	Have Done	*Am Familiar with*	*Will Learn This Year*	*No Skill; Will Not Learn This Year*
g. Select wire sizes	❑	❑	❑	❑
h. Know electrical terminology: volts, amps, watts, ohms	❑	❑	❑	❑
i. Attach wires to terminals	❑	❑	❑	❑
j. Install wire nut connectors	❑	❑	❑	❑
k. Install light fixtures	❑	❑	❑	❑
l. Install electric motors	❑	❑	❑	❑
m. Wire buildings and structures	❑	❑	❑	❑
n. Wire three-way switches	❑	❑	❑	❑
o. Wire four-way switches	❑	❑	❑	❑
13. Perform Plumbing Skills				
a. Repair leaky faucets	❑	❑	❑	❑
b. Assemble pipe and pipe fittings	❑	❑	❑	❑
c. Thread pipe	❑	❑	❑	❑
d. Measure and cut pipe	❑	❑	❑	❑
e. Cut plastic tubing/pipe	❑	❑	❑	❑
f. Install plastic tubing/pipe	❑	❑	❑	❑
g. Flare copper tubing	❑	❑	❑	❑
h. Cut copper tubing	❑	❑	❑	❑
i. Ream pipe	❑	❑	❑	❑
j. Install fixtures on plastic pipe	❑	❑	❑	❑
k. Sweat joints on copper pipe	❑	❑	❑	❑
l. Adjust air-control mechanism on a pressure system	❑	❑	❑	❑
m. Install pressure pumps	❑	❑	❑	❑
n. Repair pumps	❑	❑	❑	❑
o. Install water heaters	❑	❑	❑	❑
p. Cut soil pipe	❑	❑	❑	❑
q. Caulk joints on soil pipe	❑	❑	❑	❑
r. Lay soil pipe	❑	❑	❑	❑

Figure 20-1 Agricultural mechanics competency inventory

Instructor's Score or Approval ______________________________

Job Sheet 20-2

Name ________________________________ Date ______________

Identifying Project Selection Criteria

Objective

Upon completing this activity, you will have demonstrated the ability to identify project selection criteria.

Tools and Materials

Job Sheet 20-2

Pen or pencil

Procedure

Answer the following questions:

1. What is a project?

____

2. Why are projects used in the shop?

____

____

____

3. What are some skills a project might help you learn?

____

____

____

____

____

4. What are the recommended procedures for selecting a project?

____

____

____

____

____

____

____

____

____

____

Instructor's Score or Approval ______________________________

Job Sheet 20-3

Name ______________________________ Date ____________

Selecting a Project

Objective

Upon completing this activity, you will have demonstrated the ability to select a project.

Tools and Materials

Job Sheet 20-3

Completed competency checklist (from Class Activity 20-1)

Pen or pencil

Plan books and ideas

Procedure

1. Discuss your competency checklist results with your instructor.
2. Ask your instructor for project suggestions that will help develop new skills.
3. Search plan books for ideas.
4. Ask parents, employers, and others for ideas.
5. Look at sample projects.
6. Make a tentative decision. Briefly describe it in the space that follows:

____

____

Instructor's Score or Approval ______________________________

Job Sheet 20-4

Name ____ Date _______________

Preparing a Plan

Objective

Upon completing this activity, you will have demonstrated the ability to prepare a plan.

Tools and Materials

Job Sheet 20-4

Graph paper with ¼" squares

Pencil

Ruler

Procedure

1. Determine the longest dimension of the project.
2. Position your paper horizontally for wide projects and vertically for tall projects.
3. Divide the longest dimension of the project by 44 (or the number of squares on the paper) to find out how many feet (or inches) each square must represent.
4. Choose a scale that permits a single-view drawing with room for dimensions and notes.
5. Count the squares and draw the lines to complete one view with dimensions.
6. Complete each of the other views on separate sheets of paper.
7. Draw a plan as described in Job Sheet 18-5.
8. Review the plan with your instructor.
9. Draw better plans after obtaining the instructor's approval.
10. Complete a bill of materials.
11. Clean the area and put away materials.

Instructor's Score or Approval ____

Job Sheet 20-5

Name ______________________________ Date ____________

Identifying Design Principles

Objective

Upon completing this activity, you will have demonstrated the ability to identify basic design principles.

Tools and Materials

Job Sheet 20-5

Pen or pencil

Procedure

Complete the following sentences:

____	1. Designs should be based on	a. paint if exposed to weather.
____	2. Use extra-strength materials	b. increase strength in wood joints.
____	3. Structural steel	c. engineered plans.
____	4. Welding is stronger than	d. increase the strength of sheet-metal panels.
____	5. The most secure fastening of wood involves	e. where safety is a factor.
____	6. Bolts are more secure than	f. holds up better than wood.
____	7. Screws are more secure than	g. riveting.
____	8. Crimps and folds	h. screws.
____	9. Tongue and grooves or dowels and splines	i. gluing.
____	10. Wood and metal must be protected by	j. nails.

Instructor's Score or Approval ______________________________

SECTION 6

Tool Fitting

Unit 21

Repairing and Reconditioning Tools

Unit 22

Sharpening Tools

UNIT 21

Repairing and Reconditioning Tools

Maintaining one's tools is the sign of a good worker. Properly sharpened and conditioned tools make work easier and more accurate and save time and energy in completing tasks.

Learning how to care for tools is a valuable skill for workers who take pride in their work. Caring for tools also protects the financial investment in them.

Job Sheet 21-1

Name ______________________________ Date ____________

Restoring Leather Parts

Objective

Upon completing this activity, you will have demonstrated the ability to restore leather parts.

Tools and Materials

Job Sheet 21-1

Leather to be restored

Saddle soap

Soft cloths

Sponge

Stiff bristle brush

Soft bristle brush

Water

Procedure

1. Brush off any dirt with the stiff bristle brush.
2. Dampen the sponge or cloth with water.
3. Rub the cloth on the saddle soap until a lather forms.
4. Rub the lather onto all the leather surfaces until they are soft and clean.
5. Remove the lather with a damp cloth or sponge.
6. Allow the leather to dry.
7. Rub the leather with a soft cloth or brush.*
8. Clean the area and put away the materials.

*Leather preservative oils may be applied at this time.

Instructor's Score or Approval ______________________________

Job Sheet 21-2

Name ______________________________ Date ____________

Restoring Wooden Parts

Objective

Upon completing this activity, you will have demonstrated the ability to restore wooden parts.

Tools and Materials

Job Sheet 21-2

Wooden parts to be restored

Fine sandpaper

Soft cloths

Stiff bristle brush

Boiled linseed oil

Procedure

1. Brush away any dirt with the stiff bristle brush.
2. Sand the wooden parts to smooth rough areas.
3. Rub the wood with a soft cloth and linseed oil. (Note: Wax can be used instead of linseed oil.)
4. Repeat Step 3 as needed until the surface of the wood no longer appears to be dry.
5. Allow the wood to dry.
6. Buff with the soft, dry cloth.
7. Clean the area and put away the materials.

Instructor's Score or Approval ______________________________

Job Sheet 21-3

Name ______________________________ Date ____________

Restoring Metal Surfaces

Objective

Upon completing this activity, you will have demonstrated the ability to restore metal surfaces.

Tools and Materials

Job Sheet 21-3

Metal to be restored

Wire brush

Putty knife

Commercial solvent

Fine sandpaper

Lightweight oil

Paper towels

Soft cloths

Procedure

1. Tap the metal to remove hardened materials.
2. Use the putty knife to remove any remaining hardened materials.
3. Use the wire brush to remove pitted rust.
4. Use 400-grit sandpaper to restore a smooth surface.
5. Use the commercial solvent to clean the tool.
6. Dry the tool with paper towels or cloths.
7. Apply a thin coat of lightweight oil to smooth surfaces. Medium-weight oil may be used to coat rough surfaces.
8. Clean the area and put away the materials.

Instructor's Score or Approval ______________________________

Job Sheet 21-4

Name ______________________________________ Date ______________

Repairing Wooden Handles

Objective

Upon completing this activity, you will have demonstrated the ability to repair a wooden handle.

Tools and Materials

Job Sheet 21-4

Handle to be repaired

Water-resistant wood glue

Sandpaper

Cloths

Boiled linseed oil

Vise

Clamps

Procedure

1. Clamp the tool in the vise.
2. Force the split open and glue both sides.
3. Clamp the handle in several places.
4. Clean off excess glue.
5. Allow to dry.
6. Remove the clamps and sand the handle.
7. Treat the handle with linseed oil.
8. Polish with a clean cloth when dry.
9. Clean the area and put away the materials.

Instructor's Score or Approval ______________________________________

Job Sheet 21-5

Name ______________________________ Date ____________

Replacing Wooden Handles

Objective

Upon completing this activity, you will have demonstrated the ability to replace a wooden handle.

Tools and Materials

Job Sheet 21-5

Boiled linseed oil

Tool needing the handle

Plastic container the size of the tool

Tool handle

Wood rasp

Vise

Wooden mallet

Wooden wedge

Portable drill and bits

Metal wedges

Hacksaw

Punch

Handsaw

Safety glasses

Procedure

1. Clamp the tool head firmly in the vise.
2. Put on safety glasses.
3. Drill into the wooden remains of the old handle.
4. Use a punch to drive out the remainder of the old handle.
5. Check the new handle to see where wood needs to be removed.
6. Rasp or file the wood to a trial fit.
7. When the fit is snug, mark the handle where it meets both sides of the head.
8. Saw a kerf along the long center line of the handle. Saw approximately two-thirds of the depth of the tool head.
9. Reposition the handle in the vise and squeeze the handle until the kerf is closed.
10. Resaw the kerf.
11. Drive the handle securely into the tool head with the mallet.
12. Use the vise to grip the handle just below the head.
13. Use a hacksaw to saw the handle flush with the top of the tool head.

14. Drive in the wooden wedge until it spreads the handle and fills the head.
15. Use the hacksaw to saw the wedge flush with the handle.
16. Drive one or two steel wedges across the wooden wedge to hold it in place.
17. Place the tool head-down in a container of boiled linseed oil for several days.
18. Dry the tool and treat the handle with linseed oil.
19. Allow the handle to dry.
20. Buff the handle with the soft cloth.
21. Clean the area and put away the materials.

Instructor's Score or Approval ____

Job Sheet 21-6

Name ______________________________________ Date ______________

Reshaping a Screwdriver

Objective

Upon completing this activity, you will have demonstrated the ability to reshape a screwdriver.

Tools and Materials

Job Sheet 21-6

Screwdriver to be reshaped

File

Grinding wheel

Face shield

Leather gloves

Container of water

Protective clothing

Safety glasses

Procedure

1. Examine a similar tool to guide your shaping.
2. Put on the face shield and safety glasses.
3. Put on the protective clothing.
4. Put on the leather gloves.
5. Check the face of the grinding wheel to see that it is flat.
6. Adjust the tool rest so it is parallel to the shaft.
7. For a screwdriver or punch, square the face of the tool on the grinding wheel.
8. Taper the sides to reestablish the fit of the screwdriver in the slot on a screw.
9. Avoid excessive heat when grinding by using water to cool the tool.
10. Dress the sides of the screwdriver blade to complete the shaping.
11. Clean the area and put away the materials.

Instructor's Score or Approval ____

Job Sheet 21-7

Name ______________________________________ Date ______________

Reshaping Toolheads

Objective

Upon completing this activity, you will have demonstrated the ability to reshape a tool head.

Tools and Materials

Job Sheet 21-7

Tool head to be reshaped

Face shield

Protective clothing

Leather gloves

Grinding wheel

Container of water

Safety glasses

Procedure

1. Examine a similar tool to guide your shaping.
2. Put on the face shield and safety glasses.
3. Put on the protective clothing.
4. Put on the leather gloves.
5. Check the grinding wheel for chips or nicks.
6. Adjust the tool rest so it is parallel to the shaft.
7. Slowly grind a taper from each flat surface to the crown.
8. Avoid overheating the tool by cooling it with water.
9. Finish by twirling the tool to provide a rounded crown.
10. Clean the area and put away the materials.

Instructor's Score or Approval ______________________________________

UNIT 22

Sharpening Tools

Sharp tools are the mark of a good craftsperson. They require less energy, do finer work, and are safer for the worker. Sharpening tools are used to sharpen other tools. Learning how to sharpen tools properly enables you to do better-quality work faster. Knowing the proper angles for cutting edges is necessary to obtain a properly working tool. Keep a copy of the information in Figure 22-5 in the textbook available to help you keep your tools in proper working condition. Cutting tools use the simple lever principle of the inclined plane. Some use a single inclined plane and others a double inclined plane.

Read Unit 22 and refer to it as needed as you complete the following activities.

Class Activity 22-1: Identifying Sharpening Tools and Principles

Shop Activity 22-2: Making a Tool-Sharpening Gauge

Shop Activity 22-3: Sharpening a Center Punch

Shop Activity 22-4: Sharpening a Wood Chisel

Shop Activity 22-5: Sharpening a Knife

Shop Activity 22-6: Sharpening a Cold Chisel

Shop Activity 22-7: Sharpening an Axe or Hatchet

Shop Activity 22-8: Sharpening a Twist Bit

Shop Activity 22-9: Sharpening a Rotary Mower Blade

Shop Activity 22-10: Sharpening a Digging Tool

CAUTION!

Sharpening activities generate heat. Care must be taken to control the heat so the tool temper is preserved.

CAUTION!

Eye protection is most important in sharpening activities in order to protect individuals from flying particles.

Job Sheet 22-1

Name ______________________________________ Date _______________

Identifying Sharpening Tools and Principles

Objective

Upon completing this activity, you will have demonstrated the ability to identify sharpening tools and principles.

Tools and Materials

Job Sheet 22-1

Pen or pencil

Procedure

Using the textbook as a reference, answer the following questions:

1. Why are sharp tools safer to use?
2. What tools are used to sharpen other tools?

   ____

3. Cutting tools usually have either a single or a double cutting edge. True or false?
4. Check which tools listed in the following have single and double inclined planes as cutting edges:

Tool	*Single*	*Double*
Wood chisels	____	____
Cold chisels	____	____
Knives	____	____
Axes	____	____
Shears	____	____
Hatchets	____	____
Drills	____	____
Digging tools	____	____
Saws	____	____

5. Tools with a single inclined plane are sharpened from __________ side(s).
6. Tools with double inclined planes are sharpened from __________ side(s).
7. Edges that are convex require more force for cutting. True or false?
8. Edges that are concave require more force for cutting. True or false?
9. Saws with teeth wider than the blade are said to be __________.
10. Describe how to perform a simple test for tool sharpness.

    ____

11. What is temper?

    ____

12. Why is temper important?

    ____

13. What is whetting?

    ____

Instructor's Score or Approval ____

Job Sheet 22-2

Name ______________________________ Date ____________

Making a Tool-Sharpening Gauge

Objective

Upon completing this activity, you will have demonstrated the ability to make a useful tool-sharpening gauge.

Tools and Materials

Job Sheet 22-2

Bench rule

Drill press

26-gauge galvanized stock

Dividers

¼" twist drill

Tin snips

Flat file

Safety glasses or face shield

Hacksaw

Center punch

Scribe

Protractor

Procedure

1. Put on safety glasses before using the drill press, tin snips, hacksaw, or center punch.
2. Cut a 3" × 4½" piece of galvanized stock with all corners square.
3. Measure in 1½" from one end and mark the location.
4. Square a line across the stock through the marked location.
5. Measure 1½" along this line and lightly mark this location with the center punch.
6. Set the dividers at 1½".
7. Place one leg of the dividers in the punch mark and scribe a semicircle across the end of the stock.
8. Set the dividers to 1 3⁄16".
9. Scribe a second semicircle inside the first one.
10. Center the protractor on the scribed line and mark the inner semicircle at 20, 60, 120, and 160 degrees.
11. Use the protractor to lay out and cut triangles with a point of 20, 45, 75, and 80 degrees from heavy paper. You will use these as patterns.
12. Place the point of the pattern for the 20° triangle on the 20° mark on the inner semicircle and scribe the pattern from the inner semicircle to the outer semicircle.

13. Place the point of the pattern for the 45° triangle at the 60° mark on the inner semicircle and scribe the pattern from the inner semicircle to the outer semicircle.
14. Place the point of the pattern for the 75° triangle on the 120° mark on the inner semicircle and scribe the pattern from the inner semicircle to the outer semicircle.
15. Place the point of the 80° pattern on the 160° mark of the inner semicircle and scribe the pattern from the inner semicircle to the outer semicircle.
16. Cut along the outer semicircle with the tin snips.
17. Cut out the four scribed triangles with the tin snips. These are now angle gauges.
18. Measure 5⁄16" along the scribed line from the left side of the stock and mark the location.
19. Measure 5⁄16" from the left side at the square end of the stock and mark the location.
20. Scribe a line between these two marks.
21. Measure 9⁄16" toward the square end of the stock from the line scribed in Step 4. Mark this location on the left edge of the stock.
22. Scribe a line from this point to the point marked in Step 18.
23. Cut along this line and the line scribed in Step 20. This is now a 29° angle gauge.
24. Measure ¾" from the right along the line scribed in Step 4. Mark the location.
25. Measure ¾" from the right at the square end of the stock.
26. Scribe a line between these two locations.
27. Measure 13⁄32" down from the line scribed in Step 4. Mark this location on the line scribed in Step 26.
28. Scribe a line from this location to the right end of the line scribed in Step 4.
29. Cut out this 118° gauge.
30. Measure up ⅜" from the square end of the stock and mark the location.
31. Square a line across the stock through this location. Mark the center of this line with the center punch.
32. Drill a ¼" hole at this location.
33. Using the file, smooth all corners and edges.
34. Clean the area and store the tools.

Instructor's Score or Approval ____________________

Job Sheet 22-3

Name ______________________________ Date ____________

Sharpening a Center Punch

Objective

Upon completing this activity, you will have demonstrated the ability to sharpen a center punch.

Tools and Materials

Job Sheet 22-3

Grinding wheel

Tool gauge

A dull center punch

Safety glasses

Face shield

Procedure

1. Check the angle of the center punch with the 75° tool gauge.
2. Set the tool rest on the grinding wheel to support the punch at the correct angle.
3. Put on face shield and safety glasses.
4. Use the fingers of the left hand to hold the punch against the tool rest as a guide and twirl the punch with the right hand to obtain a sharp point.
5. Check the angle of the sharpened point with the 75° tool gauge.
6. Repeat Steps 4 and 5 until the proper point is obtained.
7. Clean the area and store the tools.

Instructor's Score or Approval ______________________________

Job Sheet 22-4

Name ______________________________ Date ____________

Sharpening a Wood Chisel

Objective

Upon completing this activity, you will have demonstrated the ability to sharpen a wood chisel.

Tools and Materials

Job Sheet 22-4

Wood chisel in need of sharpening

Grinding wheel

Oil stone and cutting oil

Tool-sharpening gauge

Try or combination square

Face shield

Safety glasses

Procedure

1. Check the angle of the wood chisel with the 29° tool gauge.
2. Set the tool rest on the grinding wheel to support the chisel at the correct angle.
3. Put on the face shield and safety glasses.
4. Hold the chisel against the tool rest and move the chisel smoothly across the wheel to obtain a smooth grind.
5. Check the squareness of the blade with the square.
6. Repeat Steps 4 and 5 until a square face is obtained.
7. Hone the chisel by placing the back of the chisel flat on the oilstone and drawing it back and forth to remove the wire edge or curl.
8. Hone the face of the chisel by placing the ground edge flat on the stone, raising the heel slightly, and moving the chisel in a figure-8 motion over the stone, using oil as needed.
9. Test the sharpness of the chisel with the 29° gauge.
10. Repeat Steps 7 and 8 until the desired sharpness is achieved.
11. Wipe the slurry from the oilstone and store the tools.

CAUTION!

Don't overheat the chisel while grinding. Use cold water to cool the chisel if necessary.

Instructor's Score or Approval ______________________________

Job Sheet 22-5

Name ______________________________ Date ____________

Sharpening a Knife

Objective

Upon completing this activity, you will have demonstrated the ability to sharpen a knife.

Tools and Materials

Job Sheet 22-5

Knife in need of sharpening

Grinding wheel

Oilstone and cutting oil

Wiping cloth

Face shield

Safety glasses

Procedure

1. Check the cutting angle of the knife with the 29° gauge.
2. Put on the face shield and safety glasses.
3. Remove any nicks by grinding.
4. Hold the knife parallel to the tool rest with the cutting edge up, and move it back and forth across the grinding wheel. Create an even grind mark about ¼" deep along the edge of the blade.
5. Use a fine grit wheel to remove the grinding marks if possible.
6. Whet the knife on the oilstone by placing the knife flat on the stone and raising the back slightly. Alternate between drawing the blade toward the cutting edge and flipping the knife over and pushing it in the other direction.
7. Test the blade's sharpness.
8. Repeat Steps 6 and 7 until the correct degree of sharpness is obtained.
9. Wipe the oilstone and store the tools.

CAUTION!

Grind slowly to avoid overheating the edge. Use cold water to cool the blade if necessary.

Instructor's Score or Approval ______________________________

Job Sheet 22-6

Name ______________________________ Date ____________

Sharpening a Cold Chisel

Objective

Upon completing this activity, you will have demonstrated the ability to sharpen a cold chisel.

Tools and Materials

Job Sheet 22-6

Cold chisel

Grinding wheel

Tool grinding gauge

Try or combination square

Face shield

Safety glasses

Procedure

1. Check the angle of the cutting edge with a 75° gauge.
2. Put on the face shield and safety glasses.
3. Adjust the tool rest on the grinding wheel to the correct angle.
4. Hold the chisel on the tool rest and move the chisel across the wheel. Afterward, flip the chisel over and repeat.
5. Check the cutting angle and the face for squareness.
6. Repeat Steps 4 and 5 until a sharp cutting edge is obtained.
7. Clean the area and store the tools.

CAUTION!

Avoid overheating the chisel.

Instructor's Score or Approval ______________________________

Job Sheet 22-7

Name ______________________________ Date ____________

Sharpening an Axe or Hatchet

Objective

Upon completing this activity, you will have demonstrated the ability to sharpen an axe or hatchet.

Tools and Materials

Job Sheet 22-7

Axe or hatchet needing sharpening

Flat file or hand stone

Grinding wheel or portable grinder

Face shield

Safety glasses

Leather gloves

Procedure (for use with a grinding wheel)

1. Check the blade angle with the 20° gauge.
2. Put on the face shield and safety glasses.
3. Put on the leather gloves.
4. Remove any nicks by grinding.
5. Draw a line parallel to the edge of the blade. This line should be between ½" to ⅝" from the blade's edge.
6. Hold the blade facing up so it forms a 10° angle with the wheel. Move it up and down as you slowly move it across the wheel with your gloved hands. This will create the wide grinding band and preserve the cutting angle.
7. Repeat the same grinding process on the opposite side of the head.
8. Repeat Steps 5 and 6 until a good edge is obtained.
9. Remove the grinding marks and wire edge with the flat file or hand stone.
10. Clean the area and store the tools.

Procedure (for use with a portable grinder)

1. Secure the blade in a machinist's vise.
2. Hold the portable grinder at the appropriate 10° angle and grind the blade to retain the proper angle.
3. Repeat the same grinding process on both sides of the head.
4. Repeat Steps 2 and 3 until a proper edge is obtained.
5. Use the hand stone or flat file to complete the edge.
6. Clean the area and store the tools.

CAUTION!

Grinding creates heat. Cool the blade as necessary with cold water to preserve the temper.

Instructor's Score or Approval ______________________________

Job Sheet 22-8

Name ______________________________ Date ______________

Sharpening a Twist Bit

Objective

Upon completing this activity, you will have demonstrated the ability to sharpen a twist bit.

Tools and Materials

Job Sheet 22-8

Twist bit in need of sharpening

Grinding wheel

Tool grinding gauge

Face shield

Safety glasses

Leather gloves

Procedure

1. Check the bit to determine
   a. if the lip fits the 118° gauge.
   b. if the heel has 10 to 12 degrees of clearance.
   c. if the dead center is the center of the bit.
2. Put on the face shield and safety glasses.
3. Put on the leather gloves.
4. Check the face of the grinding wheel for straightness, and then square to the edge and dress as needed.
5. Set the tool rest horizontal to the center of the wheel.
6. Hold the drill between your thumb and index finger with an inch exposed and with the face parallel to the face of the wheel.
7. While holding the back end of the bit, touch the lip to the wheel and lower the end of the bit while giving it a slight clockwise twist.
8. Turn the bit a half turn and repeat.
9. Repeat Step 1.
10. Repeat Steps 6 to 8 as needed.
11. Test the bit by drilling mild steel. An off-center drill will not drill an accurately sized hole.
12. Clean the area and store the tools.

CAUTION!

Avoid overheating the bit. Use cold water as needed to cool the bit.

Instructor's Score or Approval ______________________________

Job Sheet 22-9

Name ____ Date _______________

Sharpening a Rotary Mower Blade

Objective

Upon completing this activity, you will have demonstrated the ability to sharpen a rotary mower blade.

Tools and Materials

Job Sheet 22-9

Tool grinding gauge

Portable grinder

Blade in need of sharpening

Flat file or grinding wheel

Blade balancer

Face shield

Safety glasses

Leather gloves

Procedure

1. Check the cutting angle of the blade with the 45° gauge.
2. Put on the face shield and safety glasses.
3. Remove any nicks by filing or grinding. Wear the leather gloves when using the grinding wheel.
4. Secure the blade in a vise.
5. Restore the flat side of the cutting edge with the portable grinder.
6. Reverse the blade and repeat Step 5.
7. Test the blade cutting angle with the 45° gauge.
8. Test the blade balance. (Use a nail as a pivot if no balancer is available.) Grind more on the heavy side as needed.
9. Repeat Steps 5 to 8 as necessary.
10. Clean the area and store the tools.

CAUTION!

Avoid overheating the blade. Use cold water to cool it as necessary.

Instructor's Score or Approval ____

Job Sheet 22-10

Name ______________________________ Date ____________

Sharpening a Digging Tool

Objective

Upon completing this activity, you will have demonstrated the ability to sharpen a digging tool.

Tools and Materials

Job Sheet 22-10

Hand stone

Hoe, shovel, or spade needing sharpening

Flat file

Machinist's vise

Safety glasses

Ball-peen hammer

Procedure

1. Check the original cutting angle to determine which gauge to use.
2. Secure the tool in the vise.
3. Put on the safety glasses.
4. Use the hammer to straighten or remove any bends or curls.
5. Restore the cutting edge with the file.
6. Check the original cutting angle with the gauge used in Step 1.
7. Repeat Steps 5 and 6 as necessary.
8. Clean the area and put away the tools.

Instructor's Score or Approval ______________________________

SECTION 7

Gas Heating, Cutting, Brazing, and Welding

UNIT 23

Using Gas Welding Equipment

Gases that burn offer us many advantages and at the same time expose us to the danger of fire or explosion. Safety is of primary importance when dealing with compressed and flammable gases. The author of this book lived in a town where a mechanic accidentally tipped over an oxygen tank and the regulator broke off. The tank went through two concrete block walls and down an alley. Fortunately, no one was injured.

Class Activity 23-1: Identifying Safety Rules with Compressed Gases

Class Activity 23-2: Understanding the Use of Compressed Gases

Shop Activity 23-3: Setting Up the Oxyacetylene Torch

Shop Activity 23-4: Pressurizing the Torch

Shop Activity 23-5: Lighting, Adjusting, and Shutting Down the Torch

Class Activity 23-6: Identifying Oxyacetylene Flames

Job Sheet 23-1

Name ____ Date _______________

Identifying Safety Rules with Compressed Gases

Objective

Upon completing this activity, you will have demonstrated the ability to identify safety rules for compressed gases.

Tools and Materials

Job Sheet 23-1

Pen or pencil

Procedure

Complete the following worksheet:

____	1. Wear a face shield	a.	because spontaneous fire may occur.
____	2. Before using gases	b.	with a locked chain-link fence or block buildings.
____	3. Oxygen and fuel cylinders	c.	wear leather gloves and an apron.
____	4. Chain all cylinders securely	d.	any cylinder not having regulators attached.
____	5. Store cylinders	e.	at all times.
____	6. Keep caps on	f.	obtain permission from the instructor.
____	7. Follow specific procedures	g.	approved fire extinguisher in the area.
____	8. Never expose equipment to oil or grease	h.	check thoroughly for leaks.
____	9. When using gas-burning equipment, always	i.	for turning systems on or off.
____	10. Learn to recognize	j.	should point away from people and clothing.
____	11. Protect gas cylinder storage areas	k.	turn off the gas at the tank.
____	12. In case the equipment catches fire,	l.	the odors of combustible fuels.
____	13. All equipment or cylinders that can discharge gas	m.	it can be saturated with oxygen or fuel gases.
____	14. Never leave clothing where	n.	should be stored separately.
____	15. When connections are opened or cylinders changed,	o.	free from materials that will burn.
____	16. Work only in areas that are	p.	equipment connected to compressed gas cylinders.
____	17. Do not bump or put pressure on	q.	in an upright position.
____	18. Never work in an area without an	r.	outdoors or in a well-ventilated area.

Instructor's Score or Approval ____

Job Sheet 23-2

Name ______________________________ Date ____________

Understanding the Use of Compressed Gases

Objective

Upon completing this activity, you will have demonstrated an understanding of the use of compressed gases.

Tools and Materials

Job Sheet 23-2

Pen or pencil

Procedure

Answer the following questions:

1. Define welding.

____

____

2. What dangers are present when using compressed gases?

____

____

3. What does the term *flammable* mean?

____

4. What are the most common gases used for heating, cutting, and welding?

____

5. What is a manifold? Why are they used?

____

6. What is oxyacetylene?

____

____

7. Acetylene hoses are colored ____________. Oxygen hoses are colored ____________.
8. Why must the acetylene wrench be left on the cylinder at all times when it is in use?

____

____

9. Why are the threads on oxygen and acetylene hoses and fittings different?

____

10. How should gas welding equipment be tested for leaks?

____

11. Lens shade number ____________ is recommended for oxyacetylene use.

Instructor's Score or Approval ______________________________

Job Sheet 23-3

Name ______________________________________ Date _______________

Setting Up the Oxyacetylene Torch

Objective

Upon completing this activity, you will have demonstrated the ability to set up the oxyacetylene torch.

Tools and Materials

Job Sheet 23-3

Acetylene regulator

Oxygen regulator

Acetylene hose

Acetylene wrench

Oxygen hose

Oxygen tank

Acetylene tank

Torch and tips

Adjustable wrench

Oxyacetylene cart

Safety glasses

Procedure

1. Obtain permission to start the procedure.
2. Put on the safety glasses.
3. Chain the two tanks to the cart.
4. Remove the protective cap from the oxygen tank.
5. Crack the oxygen cylinder to blow out any dirt.
6. Attach the oxygen regulator to the oxygen tank.
7. Attach the green hose to the oxygen regulator gauge outlet.
8. Attach the other end of the green hose to the oxygen inlet on the torch body.
9. Open and close the acetylene tank valve quickly to blow out any dirt.
10. Attach the acetylene regulator (left-handed threads) to the acetylene tank.
11. Attach the red hose to the acetylene gauge outlet.
12. Attach the other end of the red hose to the torch body.
13. Attach the torch tip to the torch body.
14. Clean the area and put away the materials.

Instructor's Score or Approval ____

Job Sheet 23-4

Name ______________________________ Date ______________

Pressurizing the Torch

Objective

Upon completing this activity, you will have demonstrated the ability to pressurize the torch.

Tools and Materials

Job Sheet 23-4

Oxyacetylene equipment

Safety glasses

Leather gloves

Leather apron

Procedure

1. Obtain permission to start the procedure.
2. Put on the safety glasses.
3. Put on the leather apron.
4. Put on the leather gloves.
5. Close the acetylene valve on the torch.
6. Close the oxygen valve on the torch.
7. Turn the acetylene regulator handle counterclockwise until no tension is felt.
8. Turn the oxygen regulator handle counterclockwise until no tension is felt.
9. Open the oxygen cylinder valve slowly until the regulator gauge responds, and then open it fully.
10. Open the acetylene cylinder gauge slowly a half turn.
11. Open the oxygen torch valve ⅛ of a turn and rotate the regulator handle clockwise until the gauge reads 10 psi.
12. Close the oxygen torch valve.
13. Open the acetylene torch valve ⅛ of a turn and turn the acetylene regulator valve clockwise until the gauge reads 5 psi.
14. Close the acetylene torch valve.
15. Check for leaks with soapy water.
16. Clean the area and put away the materials.

Instructor's Score or Approval ______________________________

Job Sheet 23-5

Name ____ Date _______________

Lighting, Adjusting, and Shutting Down the Torch

Objective

Upon completing this activity, you will have demonstrated the ability to light, adjust, and shut down the torch.

Tools and Materials

Job Sheet 23-5

Oxyacetylene equipment

Safety glasses

Leather gloves

Leather apron

Spark lighter

Shade 5 welding face shield

Procedure

1. Obtain permission to start the procedure.
2. Put on the safety glasses.
3. Put on the shade 5 welding face shield.
4. Put on the leather apron.
5. Put on the leather gloves.
6. Pressurize the torch.
7. Open the acetylene valve ⅛ of a turn.
8. Ignite the torch with the spark lighter.
9. Open the acetylene valve slowly until the flame just touches the tip.
10. Open the oxygen valve slowly until the long inner flame matches the cone, producing a neutral flame.
11. Use more oxygen only when special circumstances require.
12. Turn off the acetylene at the cylinder.
13. Turn off the oxygen at the cylinder.
14. Open the torch valves until the regulators both return to zero.
15. Close the torch valves.
16. Turn the regulator handles counterclockwise until no tension is felt.
17. Coil the hoses and store the equipment.

Instructor's Score or Approval ____

Job Sheet 23-6

Name ______________________________ Date ____________

Identifying Oxyacetylene Flames

Objective

Upon completing this activity, you will have demonstrated the ability to identify carbonizing, neutral, and oxidizing flames.

Tools and Materials

Job Sheet 23-6

Pen or pencil

Procedure

Complete the following worksheet by identifying each type of flame:

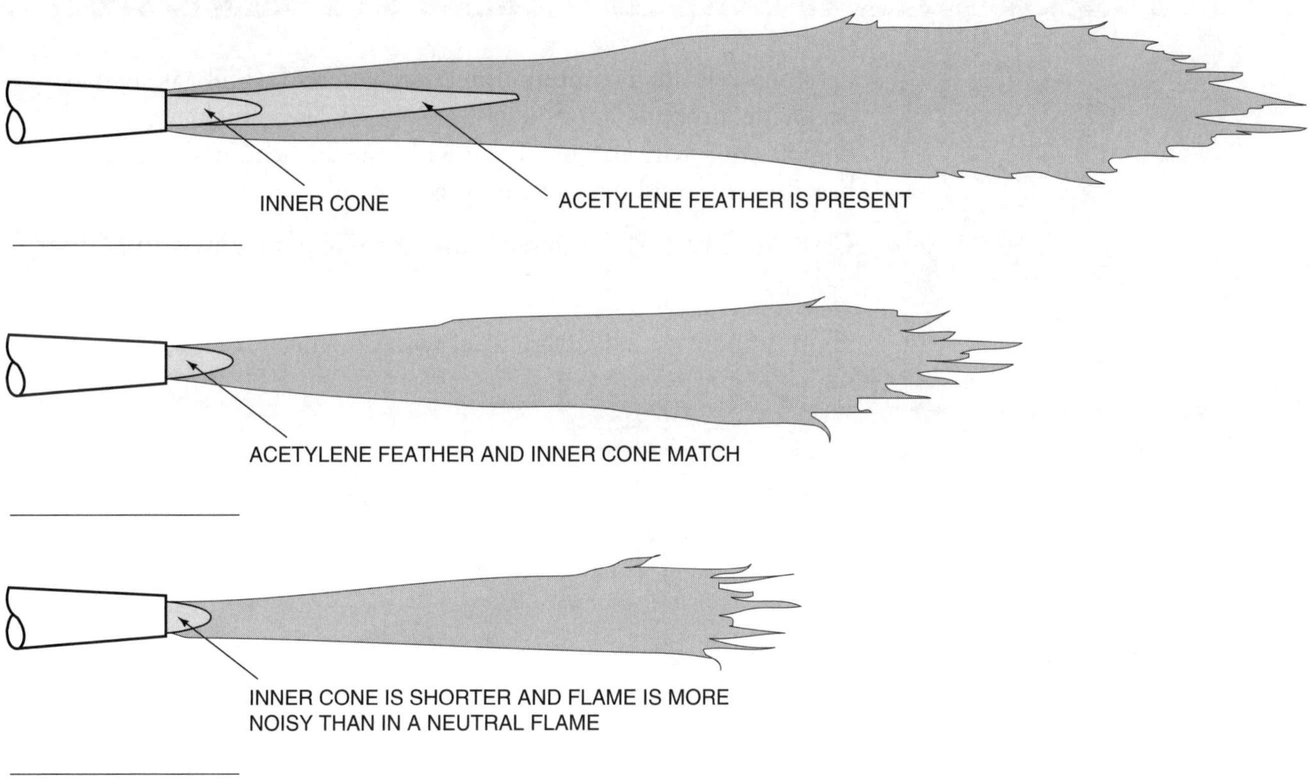

Figure 23-1 Oxyacetylene Flames

Instructor's Score or Approval ______________________________

UNIT 24

Cutting with Oxyfuels and Other Gases

Oxyfuels refer to a combination of oxygen and a fuel gas such as acetylene, propane, natural gas, or MAPP. These fuel gases, in combination with oxygen, burn with intense heat, making cutting, brazing, and welding activities possible.

Class Activity 24-1: Understanding the Use of Oxyfuels and Other Gases

Shop Activity 24-2: Cutting Steel with Oxyfuels

Job Sheet 24-1

Name ______________________________ Date ____________

Understanding the Use of Oxyfuels and Other Gases

Objective

Upon completing this activity, you will have demonstrated an understanding of the use of oxyfuels.

Tools and Materials

Job Sheet 24-1

Pen or pencil

Procedure

Answer the following questions:

1. Define the term *oxyfuel*.

____

____

2. A burning oxyfuel may reach a temperature of ______________ degrees Fahrenheit.
3. Oxyfuels are used for ______________, ______________, ______________, and ______________ metals.
4. Which fuel gas is the most commonly used fuel for welding?

____

5. List an advantage that each of the following fuel gases has over acetylene:

Natural gas ______________________________

MAPP gas ______________________________

6. Acetylene pressure must not exceed ______________ psi.
7. What is MAPP gas?

____

____

8. How do you identify when too much preheat is present?

____

9. How do you correct for too much preheat when cutting?

____

10. What is the proper torch clearance from the stock?

____

____

11. How do you identify when torch speed is too fast or slow?

____

____

12. How do you identify oxygen pressure problems?

____

____

13. What is plasma arc cutting?

____

____

Instructor's Score or Approval _______________________

Job Sheet 24-2

Name ______________________________________ Date ______________

Cutting Steel with Oxyfuels

Objective

Upon completing this activity, you will have demonstrated the ability to cut steel with oxyfuels.

Tools and Materials

Job Sheet 24-2

Leather gloves

Oxyfuel equipment

Leather apron

Cutting torch with tips

Steel to be cut

Shade 5 welding face shield

Pail of sand

Safety glasses

Procedure

1. Obtain permission to use the oxyfuel cutting outfit.
2. Put on the safety glasses.
3. Put on the shade 5 welding face shield.
4. Put on the leather apron.
5. Put on the leather gloves.
6. Clear the area of combustibles.
7. Place the pail of sand to catch the hot metal.
8. Select and install the proper tip for the job.
9. Set the working pressures for the tip used (see Figure 24-7 in the textbook).
10. Light the torch using acetylene gas only.
11. Open the valve until the flame leaves the tip, and then close it until the flame just touches the tip.
12. With the oxygen preheat valve closed, open the torch oxygen valve several turns.
13. Slowly open the preheat valve until the acetylene feathers just match the inner cone of the preheat flame.
14. Press the oxygen cutting lever and observe that the flames stay neutral.
15. Readjust to a neutral flame if necessary.
16. Hold the flame over the corner of the cut to be made.
17. Hold the inner cone about ⅛" above the metal until it turns cherry red.
18. Press the cutting lever and move slowly through the metal.
19. Shut down the torch and purge the lines.
20. Allow the metal to cool.
21. Clean the area and put away the materials.

Instructor's Score or Approval ____

UNIT 25 Brazing and Welding with Oxyacetylene

Brazing is the process of bonding with metals and alloys that melt at temperatures above 840° F. Soldering is the process of bonding with metals and alloys that melt at temperatures below 840° F. Learning these procedures will be valuable additions to your knowledge and skills.

Class Activity 25-1: Understanding Brazing and Welding with Oxyacetylene

Shop Activity 25-2: Using Flux

Shop Activity 25-3: Tinning with Solder

Shop Activity 25-4: Tinning with Braze Filler

Shop Activity 25-5: Brazing

Shop Activity 25-6: Running a Bead

Shop Activity 25-7: Braze Welding Butt Welds

Shop Activity 25-8: Braze Welding Fillet Welds

Shop Activity 25-9: Pushing the Puddle

Shop Activity 25-10: Making a Corner Weld without Filler Rod

Job Sheet 25-1

Name ______________________________ Date ____________

Understanding Brazing and Welding with Oxyacetylene

Objective

Upon completing this activity, you will have demonstrated an understanding of brazing and welding with oxy-acetylene.

Tools and Materials

Job Sheet 25-1

Pen or pencil

Procedure

Answer the following questions:

1. What do brazing, soldering, and fusion welding have in common?

____

____

2. Define the term *brazing*.

____

____

3. What is capillary action?

____

____

4. How does capillary action aid in soldering and brazing?

____

____

5. What is chemical cleaning?

____

____

6. What is a flux?

____

____

7. Why should care be taken to avoid overheating a flux?

____

____

8. How can we successfully heat two different thicknesses of metals at the same time?

____

____

9. What is the general rule used for torch flame adjustment when soldering, brazing, or welding?

____

____

10. How do you prevent burn-through when welding?

____

____

11. What do we mean by "pushing the puddle"?

____

____

____

Instructor's Score or Approval _______________________

Job Sheet 25-2

Name ______________________________ Date ____________

Using Flux

Objective

Upon completing this activity, you will have demonstrated the ability to use flux.

Tools and Materials

Job Sheet 25-2

Copper pipe, tubing, or sheet

Rosin flux

Shade 5 welding face shield

Safety glasses

Leather gloves

Steel wool or sandpaper

Leather apron

Oxyacetylene welding outfit

Procedure

1. Obtain permission to start the procedure.
2. Put on the safety glasses, welding shield, and protective clothing.
3. Set up, light, and adjust a small torch.
4. Put a small amount of rosin flux on the copper.
5. Heat the area by moving the flame in and out of the area of the flux.
6. Remove the heat when the copper suddenly becomes shiny.
7. Move the flame to an adjacent area and watch the flux flow to the new area.
8. Continue to heat the copper until the heat burns the flux and blackens the area. This demonstrates what happens when too much heat is applied.
9. Remove the burned flux with sandpaper or steel wool.
10. Reapply the flux and reheat.*
11. Clean the area and put away the materials.

*Note that the fluxing action can only be repeated after the mechanical cleaning of the area. The ability to control the fluxing action is necessary for successful bonding of solder or brazing materials.

Instructor's Score or Approval ______________________________

Job Sheet 25-3

Name ______________________________________ Date ______________

Tinning with Solder

Objective

Upon completing this activity, you will have demonstrated the ability to tin with solder.

Tools and Materials

Job Sheet 25-3

A piece of copper

Oxyfuel torch

Shade 5 welding face shield

Safety glasses

Leather gloves

Leather apron

Solder

Fine emery cloth

Paste flux

Damp cloth

Procedure

1. Obtain permission to start the procedure.
2. Put on the safety glasses, welding shield, and protective clothing.
3. Clean the copper with the emery cloth.
4. Light and adjust an oxyfuel or propane torch.
5. Heat the copper until the fluxing action starts.
6. Touch the solder to the metal and feed it down onto the copper as it is tinned.
7. Continue to move the heat until the area is tinned.
8. If needed, smooth and even the solder with the flame.
9. Wipe the area quickly with the damp cloth to remove excess solder and improve the appearance.
10. Clean the area and put away the materials.

Instructor's Score or Approval ______________________________________

Job Sheet 25-4

Name ______________________________ Date ____________

Tinning with Braze Filler

Objective

Upon completing this activity, you will have demonstrated the ability to tin with braze filler.

Tools and Materials

Job Sheet 25-4

Oxyfuel outfit

Braze filler rod

Flux

Safety glasses

Leather gloves

Leather apron

Shade 5 welding face shield

Fine emery cloth

Firebricks

Nonrusted steel plate approximately ⅛" × 2" × 6"

Procedure

1. Obtain permission to start the procedure.
2. Put on the safety glasses, welding shield, and protective clothing.
3. Clean the steel with the emery cloth.
4. Place the steel on the firebricks.
5. Light and adjust the torch using a small tip.
6. Heat the steel until the center is a dull red.
7. As you heat the base metal, heat the filler rod by holding it in the flame.
8. Flux the heated rod by dipping it in the flux can.
9. Place the flux-coated rod in the flame and touch the base metal.
10. When the temperature is correct, fluxing action takes place, cleaning impurities.
11. As fluxing action occurs, watch for the filler rod to melt, playing the flame over the metal, the rod, and flux.
12. Control the temperature to tin the whole area.
13. Clean the area and put away the materials.

Instructor's Score or Approval ______________________________

Job Sheet 25-5

Name ______________________________________ Date ______________

Brazing

Objective

Upon completing this activity, you will have demonstrated the ability to braze.

Tools and Materials

Job Sheet 25-5

Three pieces of flat steel, ⅛" × 2" × 2"

Emery cloth

Safety glasses

Leather gloves

Leather apron

Shade 5 welding face shield

Brazing rod

Flux

Oxyfuel torch

Procedure

1. Obtain permission to start the procedure.
2. Put on the safety glasses, welding shield, and protective clothing.
3. Clean 1" of the edge on each of two pieces.
4. Lay the two steel pieces on the firebricks so that clean ½" areas overlap, using the third piece to support the second (refer to Figure 25-10 in the textbook).
5. Clamp the assembly or hold it in place with a brick.
6. Light and adjust the torch.
7. Apply heat until both pieces are a dull red.
8. Add flux by way of the rod.
9. Continue to heat the metal to a dull red and move the rod in and out of the flame to deposit metal.
10. Direct most of the heat on the upper piece and stay back from the edge to draw the filler into the joint.
11. Allow the assembly to cool.
12. Turn the assembly over and check for completeness.
13. Clean the area and put away the materials.

Instructor's Score or Approval ______________________________________

Job Sheet 25-6

Name ______________________________ Date ____________

Running a Bead

Objective

Upon completing this activity, you will have demonstrated the ability to run a bead.

Tools and Materials

Job Sheet 25-6

Oxyfuel torch

Shade 5 welding face shield

Safety glasses

Leather gloves

Leather apron

Clean ⅛" × 2" × 6" steel

Flux

Firebricks

Emery cloth

Filler rod

Procedure

1. Obtain permission to start the procedure.
2. Put on the safety glasses, welding shield, and protective clothing.
3. Clean the metal with the emery cloth.
4. Put the metal on a firebrick.
5. Light and adjust the torch.
6. Heat the metal ½" from the end (you will run the bead across the metal).
7. Heat the rod and dip it in the flux.
8. Touch the fluxed rod to the metal and, as it melts, move it in and out of the flame to lay the bead.
9. The bead is formed by depositing the metal along a fluxed path across the metal.
10. Lay additional beads at 1" intervals until you can lay complete beads that are straight and even.
11. Clean the area and put away the materials.

Instructor's Score or Approval ______________________________

Job Sheet 25-7

Name ______________________________ Date ____________

Braze Welding Butt Welds

Objective

Upon completing this activity, you will have demonstrated the ability to braze weld a butt joint.

Tools and Materials

Job Sheet 25-7

Two pieces of 16-gauge steel 2" × 2"

Oxyfuel torch

Filler rod

Flux

Fine emery cloth

Shade 5 welding face shield

Safety glasses

Leather gloves

Leather apron

Firebricks

Procedure

1. Obtain permission to start the procedure.
2. Put on the safety glasses, welding shield, and protective clothing.
3. Clean pieces of the metal with the emery cloth.
4. Place them on the flat firebrick with a 1/32" gap between them.
5. Weight the two pieces with firebricks to hold them in position.
6. Light and adjust the torch.
7. Heat a spot at one end of the joint, flux the area, and deposit a spot of filler material. This is a tack weld.
8. Place a second tack at the other end of the joint.
9. Lay a bead over the entire length of the joint.
10. Clean the area and put away the materials.

Instructor's Score or Approval ______________________________

Job Sheet 25-8

Name ____ Date ______________

Braze Welding Fillet Welds

Objective

Upon completing this activity, you will have demonstrated the ability to braze weld a fillet joint.

Tools and Materials

Job Sheet 25-8

Oxyfuel torch

Firebricks

Emery cloth

Two pieces of 16-gauge steel 1½" × 6"

Shade 5 welding face shield

Safety glasses

Leather gloves

Leather apron

Filler rod

Flux

Procedure

1. Obtain permission to start the procedure.
2. Put on the safety glasses, welding shield, and protective clothing.
3. Clean the metal with the emery cloth.
4. Place one piece flat on the firebrick.
5. Place the second piece on edge on top of the first. Use the firebrick to prop them in this position. The position should be in the form of an L or a T to create a 90° angle.
6. Light and adjust the torch.
7. Place a tack weld at each end and the middle.
8. Lay a bead along the length of the joint.
9. Clean the area and put away the materials.

Instructor's Score or Approval ____

Job Sheet 25-9

Name ______________________________ Date ____________

Pushing the Puddle

Objective

Upon completing this activity, you will have demonstrated the ability to push the puddle.

Tools and Materials

Job Sheet 25-9

Oxyfuel torch

Firebricks

Shade 5 welding face shield

Safety glasses

Leather gloves

Leather apron

16-gauge 2" × 6" steel

Emery cloth

Procedure

1. Obtain permission to start the procedure.
2. Put on the safety glasses, welding shield, and protective clothing.
3. Clean the metal with the emery cloth.
4. Weight the metal on a firebrick.
5. Light and adjust the torch.
6. Start at one end and point the tip at a 45° angle in the direction of travel, with a gap of ⅛" between tip and metal.
7. When the metal starts to melt, move the torch in a small circular pattern to form the bead.*
8. Lift the torch slowly at the end of the bead.
9. Repeat beads until the plate is full.
10. Clean the area and put away the materials.

*Examine Figure 25-15 in the textbook to compare your beads.

Instructor's Score or Approval ______________________________

Job Sheet 25-10

Name ____ Date _______________

Making a Corner Weld without Filler Rod

Objective

Upon completing this activity, you will have demonstrated the ability to make a corner weld without filler rod.

Tools and Materials

Job Sheet 25-10

Oxyfuel torch

Shade 5 welding face shield

Safety glasses

Leather gloves

Leather apron

Emery cloth

Two pieces of steel 16-gauge 1½" × 6"

Firebricks

Procedure

1. Obtain permission to start the procedure.
2. Put on the safety glasses, welding shield, and protective clothing.
3. Clean the edges of the metal with the emery cloth.
4. Put the two together so their edges make a 60° tent.
5. Light and adjust the torch.
6. Tack weld the ends and two locations in the middle.
7. Make a bead all along the joint, allowing the edges of each piece to melt together to form the puddle.
8. Allow the workpiece to cool.
9. Clean the area and put away the materials.

Instructor's Score or Approval ____

SECTION 8

Arc Welding

Unit 26

Selecting and Using Arc Welding Equipment

Unit 27

Arc Welding Mild Steel and GMAW/ GTAW Welding

UNIT 26 Selecting and Using Arc Welding Equipment

Arc welding equipment is widely used in agricultural construction and maintenance. Knowledge of arc welding and skill in its use will be valuable to you.

Class Activity 26-1: Understanding Arc Welding

Class Activity 26-2: Identifying Arc Welding Advantages

Class Activity 26-3: Identifying Arc Welding Safety Items

Class Activity 26-4: Identifying Electrode Code Numbers

Job Sheet 26-1

Name ______________________________ Date ____________

Understanding Arc Welding

Objective

Upon completing this activity, you will have demonstrated an understanding of arc welding.

Tools and Materials

Job Sheet 26-1

Pen or pencil

Procedure

Answer the following questions:

1. What are the benefits of slag formation in welding?
____
____
2. The arc temperature can reach ____________________ degrees Fahrenheit.
3. What factors can be manipulated to change this temperature?
____
____
4. What does the term *duty cycle* mean when used to describe arc welders?
____
____
5. What are the differences between direct current welders and alternating current welders?
____
____
____
6. What are the common amperages of popular agricultural welders?
____
7. Portable DC welders are commonly driven by ____________________.
8. Portable welders may also be used as ____________________.
9. What happens if too-small welding cables are used?
____
____
10. Describe a desirable electrode holder.
____
____
11. What is a ground clamp?
____
____
____

Instructor's Score or Approval ______________________________

Job Sheet 26-2

Name ______________________________________ Date ______________

Identifying Arc Welding Advantages

Objective

Upon completing this activity, you will have demonstrated the ability to identify the advantages of arc welding.

Tools and Materials

Job Sheet 26-2

Pen or pencil

Procedure

Match the items in Column I with those in Column II.

Column I	*Column II*
____ 1. Electricity	a. portable welders.
____ 2. Some arc welders can be used on	b. fast and reliable.
____ 3. Brazing and heating as well as welding	c. an inexpensive source of heat for welding.
____ 4. Engine-driven	d. can be learned quickly.
____ 5. Arc welding is	e. ordinary 230-volt household or farmstead wiring.
____ 6. Farm arc welders are	f. relatively inexpensive.

Instructor's Score or Approval ____

Job Sheet 26-3

Name ______________________________________ Date ______________

Identifying Arc Welding Safety Items

Objective

Upon completing this activity, you will have demonstrated the ability to identify arc welding safety items.

Tools and Materials

Job Sheet 26-3

Pen or pencil

Procedure

Match the items in Column I with those in Column II.

Column I		*Column II*
____	1. Can convert high-voltage/low-amperage to low-voltage/high-amperage	a. DC welders
____	2. Driven by electric motors or gasoline or diesel engines	b. AC welders
____	3. Large cables can carry	c. electrode holder
____	4. Welding curtains serve as	d. high amperage
____	5. Spring-loaded device	e. protection for others
____	6. Welding tables should be	f. comfortable workstations

Instructor's Score or Approval ______________________________________

Job Sheet 26-4

Name ____ Date _______________

Identifying Electrode Code Numbers

Objective

Upon completing this activity, you will have demonstrated the ability to identify the electrode code numbers.

Tools and Materials

Job Sheet 26-4

Pen or pencil

Procedure

Match the code number listed on the left with its meaning on the right. Meanings can be used more than once.

Remember: The answer pertains only to the number or letter, not to the blanks.

	Code Number	*Meaning*
____	1. E _ _ _ _	a. thousand pounds of tensile strength
____	2. _ _ _ _ 1	b. usable in all directions
____	3. _ _ _ 1 _	c. usable only in the flat or horizontal positions
____	4. _ _ _ _ 2	d. usable for vertical down only
____	5. _ _ _ _ 3	e. DC reverse polarity only
____	6. _ _ _ _ 4	f. AC and DC reverse polarity
____	7. _ _ _ _ 5	g. AC and DC straight polarity
____	8. _ _ _ _ 6	h. AC and DC
____	9. _ _ _ _ 8	i. DC reverse polarity
____	10. _ _ _ 2 _	j. electrode
____	11. _ _ _ 4 _	
____	12. _ 7 0 _ _	
____	13. _ _ _ _ 0	

Instructor's Score or Approval ____

UNIT 27

Arc Welding Mild Steel and GMAW/GTAW Welding

Practice in arc welding can quickly develop skill. In this unit, you practice arc welding skills. Because the use of arc welders presents a fire danger from the intense heat, you should practice all safety precautions.

Class Activity 27-1: Understanding Arc Welding of Mild Steel

Shop Activity 27-2: Striking the Arc

Shop Activity 27-3: Making a Practice Bead

Shop Activity 27-4: Making a Butt Weld

Shop Activity 27-5: Checking the Butt Weld

Shop Activity 27-6: Making the Fillet Weld

Shop Activity 27-7: Making a Horizontal Weld

Shop Activity 27-8: Making a Vertical Weld

Shop Activity 27-9: Making an Overhead Weld

Shop Activity 27-10: Piercing with the Arc Welder

Shop Activity 27-11: Cutting with the Arc Welder

Class Activity 27-12: Identifying the Components of a GMAW Welder

Class Activity 27-13: Identifying GMAW Welding Defects

Job Sheet 27-1

Name ______________________ Date ____________

Understanding Arc Welding of Mild Steel

Objective

Upon completing this activity, you will have demonstrated an understanding of arc welding with mild steel.

Tools and Materials

Job Sheet 27-1

Pen or pencil

Procedure

Answer the following questions:

1. Why should the welding area have metal benches?

____

____

2. The fire extinguisher in the area should be type ____________.
3. List the personal protection gear that should be in the welding area.

____

____

____

4. Shade number ____________ lenses should be used in arc welding.
5. The recommended electrode angle is ____________ degrees.
6. Why is sound a good guide to arc length?

____

____

____

7. What are the best indicators of proper speed of travel?

____

____

____

8. What happens when the amperage is set too low?

____

____

____

9. What happens when the amperage is set too high?

____

____

____

Instructor's Score or Approval ______________________

Job Sheet 27-2

Name ______________________________ Date ____________

Striking the Arc

Objective

Upon completing this activity, you will have demonstrated the ability to strike an arc.

Tools and Materials

Job Sheet 27-2

Arc welder

E6011 electrodes

Chipping hammer

Welding helmet with shade 10 lens

Welding table

Mild steel scrap or ⅛" to ¼" × 2" × 6"

Leather gloves

Leather apron

Safety glasses

Procedure

1. Obtain permission to use the arc welder.
2. Put on the welding helmet, safety glasses, leather gloves, and apron.
3. Set the arc welder to 80–100 amps.
4. Clamp the metal to be used.
5. Put an electrode in the electrode holder.
6. Turn on the welder.
7. Lower the helmet into position.
8. Lower the electrode to touch the metal and then lift it quickly, as if striking a match.
9. If you are too slow in lifting and the electrode sticks, twist it free or release the holder.
10. If the coating on the electrode is damaged, put it aside and use another (this damaged electrode can be used later as you gain more welding experience).
11. If you lift too high and the arc is broken, try again and stay closer to the metal.
12. Feed the electrode down slowly as it burns away. Travel forward 1" and break the arc.
13. Practice Steps 8 to 12 until you have sixteen 1" beads on your metal.
14. Compare your beads with Figure 27-10 in the textbook.
15. Clean the area and put away the materials.

Instructor's Score or Approval ______________________________

Job Sheet 27-3

Name ______________________________ Date ____________

Making a Practice Bead

Objective

Upon completing this activity, you will have demonstrated the ability to run a bead.

Tools and Materials

Job Sheet 27-3

Arc welder

E6011 electrodes

Welding helmet with shade 10 lens

Leather gloves

Leather apron

Welding table

Chipping hammer

Mild steel ¼" × 4" × 6"

Welding clamp

Face shield, unless the helmet is equipped with a flip-up lens

Safety glasses

Procedure

1. Obtain permission to use the arc welder.
2. Put on the welding helmet, safety glasses, leather gloves, and apron.
3. Clamp the metal to the welding table.
4. Set the amperage to 110 amps.
5. Put the electrode in the holder.
6. Turn on the welder and lower the helmet.
7. Lean the electrode about 80° and strike the arc.
8. Move quickly to the starting spot at the edge of the plate.
9. Feed the electrode down slowly, watching the puddle and moving forward.
10. At the end of the bead, lift the electrode slowly until the center of the puddle fills and the arc goes out.
11. Clean the bead with the chipping hammer using the flip-up clear lens or a face shield.
12. Compare your bead with those in Figure 27-14 in the textbook .
13. Repeat this procedure until you can consistently make quality beads.
14. Lower the amperage if the pad becomes too hot.
15. Clean the area and put away the materials.

Instructor's Score or Approval ______________________________

Job Sheet 27-4

Name ____ Date _______________

Making a Butt Weld

Objective

Upon completing this activity, you will have demonstrated the ability to make a butt weld.

Tools and Materials

Job Sheet 27-4
Arc welder
Welding table
Chipping hammer
Welding helmet with shade 10 lens
Grinder
Leather gloves
Leather apron
Two pieces of mild steel ¼" × 2" × 6"
E6011 electrodes
Face shield unless the helmet has a flip-up lens
Safety glasses

Procedure

1. Obtain permission to use the arc welder.
2. Put on the welding helmet, safety glasses, leather gloves, and apron.
3. Grind one edge of each piece down to ⅛" thick.
4. Place the pieces ⅛" apart.
5. Set the amperage to 110.
6. Turn on the welder and lower the helmet.
7. Strike the arc and move quickly to tack each end.
8. Flatten the pieces if needed.
9. Create an even ¼"-wide bead.
10. Chip the slag with the chipping hammer.
11. Examine the weld. The answer to all of the following questions should be yes.
    a. Do the sides blend in evenly with the base metal?
    b. Are the semicircles of the bead evenly spaced?
    c. Does the weld go all the way through the metal?
    d. Does the weld start at the starting edge?
    e. Does the weld fill the groove at the starting edge?
    f. Does the weld go all the way to the finishing edge?
    g. Does the weld fill the groove at the finishing edge?
12. Clean the area and put away the materials.

Instructor's Score or Approval ____

Job Sheet 27-5

Name ________________________ Date ____________

Checking the Butt Weld

Objective

Upon completing this activity, you will have demonstrated the ability to evaluate a butt weld.

Tools and Materials

Job Sheet 27-5

Leather gloves

Leather apron

Face shield

Hacksaw

Machinist's vise

Two-pound hammer

Safety glasses

Procedure

1. Saw 1" off the end of the butt weld completed for Job Sheet 27-4.
2. Examine the cross section for voids or slag.
3. Repeat Steps 1 and 2.
4. Clamp the remainder in a heavy vise.
5. With a heavy hammer, attempt to break the weld.
6. If the weld breaks, determine the problem by comparing your weld with the text and conferring with your instructor.
7. Repeat the practice as needed to evaluate your welds.
8. Clean the area and put away the materials.

Instructor's Score or Approval ________________________

Job Sheet 27-6

Name ______________________________ Date ____________

Making the Fillet Weld

Objective

Upon completing this activity, you will have demonstrated the ability to make a fillet weld.

Tools and Materials

Job Sheet 27-6

Arc welder

Welding helmet with shade 10 lens

Chipping hammer

Leather gloves

Leather apron

E6011 electrodes

Face shield

Two pieces of steel ⅛" × 1" × 6"

Welding clamp

Safety glasses

Procedure

1. Obtain permission to use the arc welder.
2. Put on the protective equipment.
3. Set one piece of metal on the table and clamp another piece on top of it vertically.
4. Set the welder to 110 amps.
5. Tack weld both ends.
6. Square the piece to 90°.
7. Hold the electrode at 45° and lean it 15° in the direction of travel.
8. Run the bead all the way along the joint.
9. Maneuver the tip of the electrode to keep the penetration equal.
10. Run a bead on the other side, alternating passes to control distortion.
11. Chip and examine the weld.
12. Clean the area and put away the materials.

Instructor's Score or Approval ______________________________

Job Sheet 27-7

Name ______________________________________ Date ______________

Making a Horizontal Weld

Objective

Upon completing this activity, you will have demonstrated the ability to make a horizontal weld.

Tools and Materials

Job Sheet 27-7

Arc welder

E6011 electrodes

⅛" or ¼" × 4" × 6" practice pad

Chipping hammer

Welding helmet with shade 10 lens

Leather gloves

Leather arm and shoulder protection

Leather apron

Welding table

Face shield

Safety glasses

Procedure

1. Obtain permission to use the arc welder.
2. Put on the protective equipment.
3. Set the welder to 110 amps.
4. Weld two small pieces to the pad so it can be set on edge and clamped to the table.
5. Sit in front of the table.
6. Start at the left end with the electrode tilted at 70° to the plate in the direction of travel. The electrode should also be tilted so it tips upward at about 80° to the plate.
7. Move the electrode at a speed that creates a proper bead without sagging or undercutting.
8. If necessary, adjust the amperage and run more beads to develop the skill.
9. Clean the area and put away the materials.

Instructor's Score or Approval ______________________________________

Job Sheet 27-8

Name ______________________________ Date ____________

Making a Vertical Weld

Objective

Upon completing this activity, you will have demonstrated the ability to make a vertical weld.

Tools and Materials

Job Sheet 27-8

Arc welder

Welding table

Clamps

E6011 electrodes

Welding helmet with shade 10 lens

Chipping hammer

Leather gloves

Leather apron

Leather shoulder and arm protection

Face shield

⅛" or ¼" × 4" × 6" practice pad

Safety glasses

Procedure

1. Obtain permission to use the arc welder.
2. Put on the protective equipment.
3. Set up the practice pad as in Job Sheet 27-7.
4. Strike an arc at the bottom and work up.
5. Use a wide weave pattern to form the bead.
6. If the weld is too hot, stop and clean the weld. Readjust the amperage until control is achieved.
7. Practice until you can make a sound bead.
8. For a vertical-up fillet weld, use a T-pattern weave from one side to the other.
9. Clean the area and put away the materials.

Instructor's Score or Approval ______________________________

Job Sheet 27-9

Name ______________________________ Date ____________

Making an Overhead Weld

Objective

Upon completing this activity, you will have demonstrated the ability to make an overhead weld.

Tools and Materials

Job Sheet 27-9

Arc welder

Helmet with shade 10 lens

Leather arm and shoulder protection

Leather gloves

Leather apron

E6011 electrodes

Practice pad

Clamps

Welding table

Face shield

Safety glasses

Procedure

1. Obtain permission to use the arc welder.
2. Clamp the pad in an overhead position.
3. Put on the protective equipment.
4. Set the amperage to 100.
5. Hold the electrode nearly straight up.
6. Run the bead, adjusting amperage as necessary.
7. Practice until you develop the skill.
8. Clean the area and put away the materials.

CAUTION!

When first learning to weld in the overhead position, do not stand directly under the weld.

Instructor's Score or Approval ______________________________

Job Sheet 27-10

Name ______________________________________ Date ______________

Piercing with the Arc Welder

Objective

Upon completing this activity, you will have demonstrated the ability to pierce with the arc welder.

Tools and Materials

Job Sheet 27-10

Arc welder

Helmet with shade 10 lens

E6011 electrodes

Welding table

Clamps

Metal to pierce

Bucket of sand

Leather gloves

Leather apron

Leather foot and leg protection

Safety glasses

Procedure

1. Obtain permission to use the arc welder.
2. Clamp the metal so it extends 4" over the edge of the table.
3. Place the bucket of sand to collect the hot metal.
4. Put on the protective equipment.
5. Set the welder to 150 amperes.
6. Strike and hold a long arc (¼–⅛" arc) until the metal turns bright red and melts.
7. Rotate the electrode to create an enlarged molten puddle.
8. Quickly push the electrode into and through the puddle.
9. Move the electrode around the edge of the hole to enlarge it as needed.
10. If the hole must be round, put it over the hole in an anvil and drive a punch through the hot metal.
11. Smooth the underside with the grinder.
12. Clean the area and put away the materials.

Instructor's Score or Approval ______________________________________

Job Sheet 27-11

Name ________________________________ Date ____________

Cutting with the Arc Welder

Objective

Upon completing this activity, you will have demonstrated the ability to cut with the arc welder.

Tools and Materials

Job Sheet 27-11

Arc welder

Helmet with shade 10 lens

Welding table

Bucket of sand

Clamps

Metal to be cut

E6011 electrodes

Grinder

Leather gloves

Leather apron

Leather foot and leg protection

Safety glasses

Procedure

1. Obtain permission to use the arc welder.
2. Clamp the metal so it extends over the edge of the table.
3. Position the sand to catch the hot metal.
4. Put on the protective equipment.
5. Set the welder to 150 amperes.
6. Hold a long arc at the edge of the metal until it is molten.
7. Use quick up-and-down motions to chop the metal away and create a kerf across the metal.
8. Smooth with the grinder as needed.
9. Clean the area and put away the materials.

Instructor's Score or Approval ________________________________

Job Sheet 27-12

Name ________________________________ Date ______________

Identifying the Components of a GMAW Welder

Objective

Upon completing this activity, you will have demonstrated the ability to identify the components of the GMAW welder.

Tools and Materials

Job Sheet 27-12

Pen or pencil

Procedure

Complete the following worksheet by identifying the components of the GMAW welder.

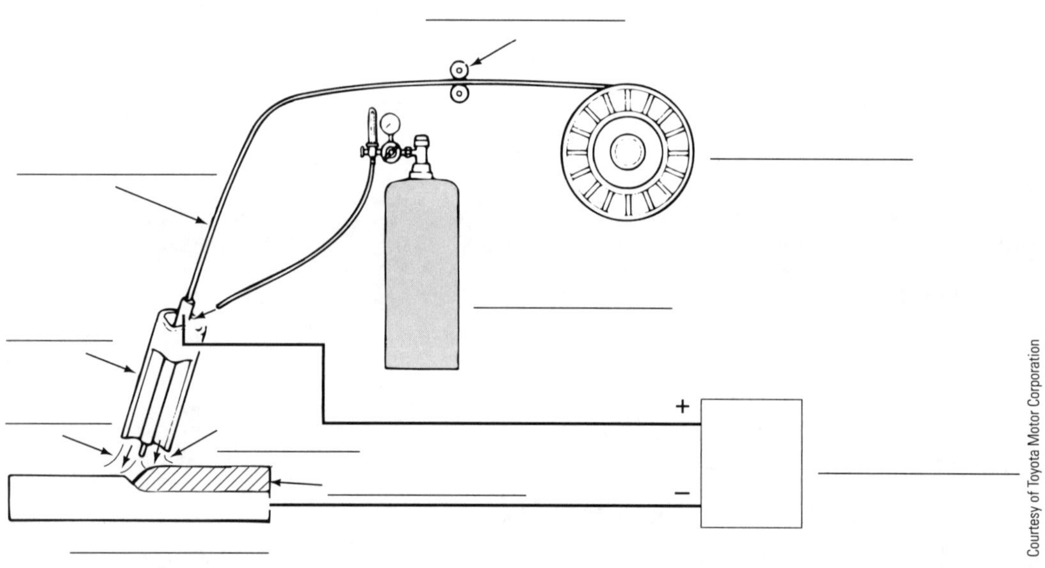

Figure 27-1 Components of a GMAW Welder

Instructor's Score or Approval ________________________________

Job Sheet 27-13

Name ______________________________ Date ____________

Identifying GMAW Welding Defects

Objective

Upon completing this activity, you will have demonstrated the ability to identify GMAW welding defects.

Tools and Materials

Job Sheet 27-13

Pen or pencil

Procedure

Complete the following worksheet:

Defect Condition	Causes
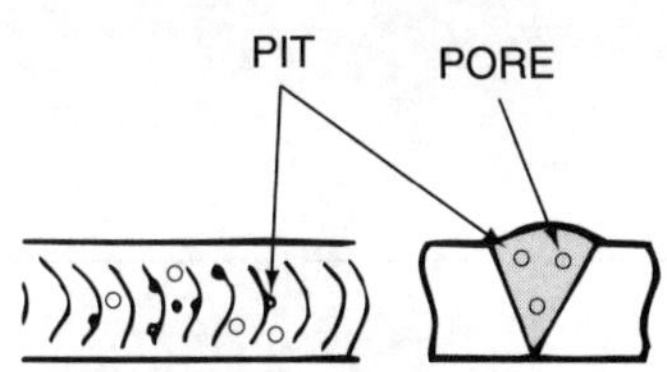	
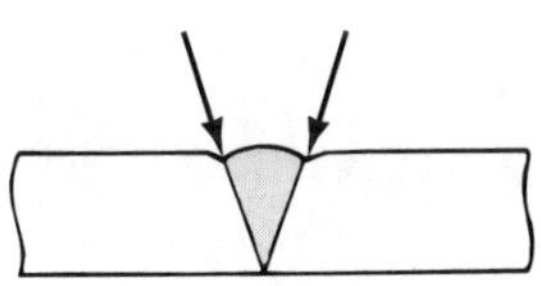	
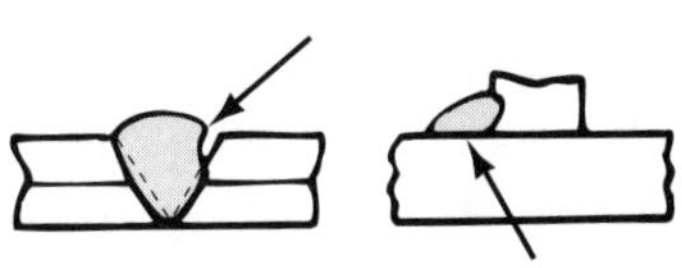	
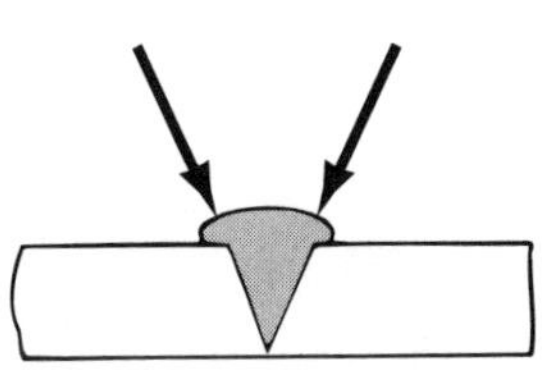	

Figure 27-2 GMAW Welding Defects (*continued*)

Instructor's Score or Approval ______________________________

Defect Condition	Causes
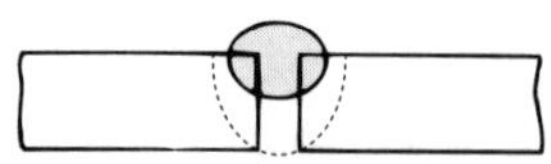	
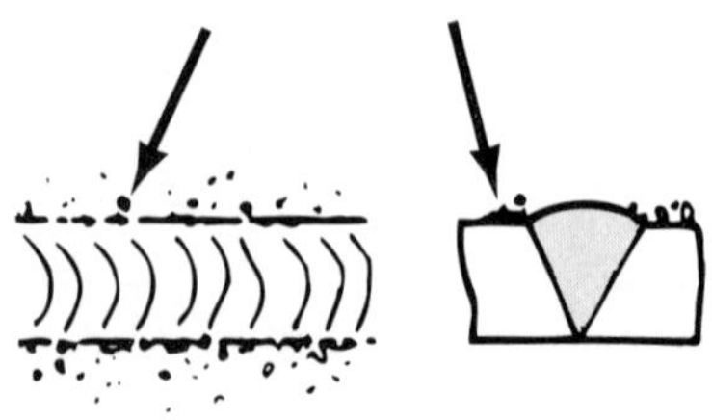	

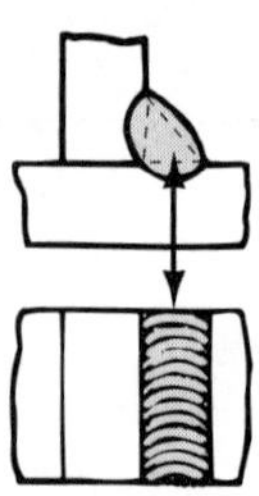	
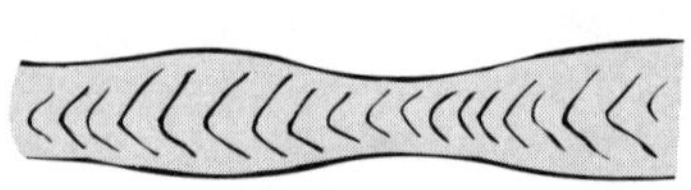	
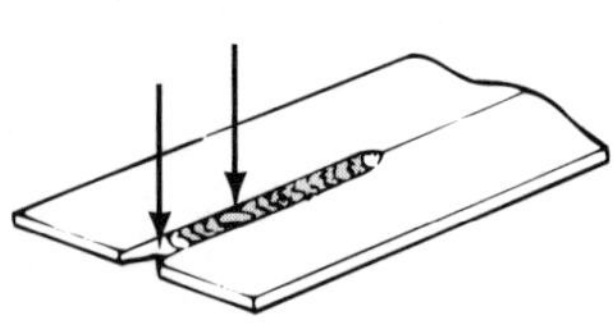	

Figure 27-2 GMAW Welding Defects

Instructor's Score or Approval ______________________________

SECTION 9

Painting

UNIT 28

Preparing Wood and Metal for Painting

Most agricultural machinery and buildings are exposed to the weather some or all of the time. If metal and wood are exposed to the weather, they need to be protected. The most common weather protection is a good coat of paint. In this unit, you learn how to prepare wood and metal for painting.

Class Activity 28-1: Understanding Wood and Metal Preparation

Shop Activity 28-2: Preparing New Wood

Shop Activity 28-3: Preparing Old Wood

Shop Activity 28-4: Steam or Pressure Cleaning Machinery

Shop Activity 28-5: Preparing New Metal for Painting

Shop Activity 28-6: Preparing Previously Painted Metal

Shop Activity 28-7: Masking for Spray Painting

Class Activity 28-8: Estimating Paint Jobs

Job Sheet 28-1

Name ________________________________ Date ____________

Understand Wood and Metal Preparation

Objective

Upon completing this activity, you will have demonstrated an understanding of wood and metal preparation for painting.

Tools and Materials

Job Sheet 28-1

Pen or pencil

Procedure

Answer the following questions:

1. What are the primary materials used in agricultural buildings and equipment?
____
____
2. What is the most common protection from moisture damage?
____
____
3. Define paint.
____
____
4. Define preservatives.
____
____
5. What does the term *new wood* refer to?
____
____
6. What is the purpose of a sealer?
____
____
____
7. Why should you wear a respirator when removing old paint?
____
____
8. Define primer.
____
____

9. Define caulk.

____

____

10. How are caulk and primer used?

____

____

Instructor's Score or Approval _______________________

Job Sheet 28-2

Name ______________________________ Date ____________

Preparing New Wood

Objective

Upon completing this activity, you will have demonstrated the ability to prepare new wood for painting.

Tools and Materials

Job Sheet 28-2

New wood to be painted

Stiff bristle brush

Mineral spirits

Rags

Sandpaper

Household ammonia

Procedure

1. Brush any dirt off the wood.
2. Clean off any grease or oil with a rag dipped in mineral spirits.
3. Clean off any wax with household ammonia.
4. Sand to remove any rough spots or to remove dirt not dislodged by the brush.
5. Allow the wood to dry.
6. Clean the area and put away the materials.

Instructor's Score or Approval ______________________________

Job Sheet 28-3

Name ______________________________ Date ____________

Preparing Old Wood

Objective

Upon completing this activity, you will have demonstrated the ability to prepare old wood for painting.

Tools and Materials

Job Sheet 28-3

Old wood to be painted

Paint scrapers

Wire brush

Paint and varnish remover

Mineral spirits

Rags

Sandpaper

Putty or plastic wood

Respirator mask

Procedure

1. Remove loose paint with a scraper or wire brush. Remember to wear a respirator if the paint being removed might contain lead.
2. Remove grease or oil with a rag soaked in mineral spirits.
3. Use paint and varnish remover if a smooth surface is needed.
4. Sand spots needing special attention. Remember to wear a respirator if the paint being removed might contain lead.
5. Fill all holes with putty or plastic wood.
6. Sand the putty or plastic wood when dry.
7. Clean the area and put away the materials.

Instructor's Score or Approval ______________________________

Job Sheet 28-4

Name ____ Date _______________

Steam or Pressure Cleaning Machinery

Objective

Upon completing this activity, you will have demonstrated the ability to use a steam cleaner or pressure washer.

Tools and Materials

Job Sheet 28-4

Steam or pressure cleaner

Machine to be cleaned

Cleaning compound

Water supply

Plastic bags

Masking tape

Fuel for the cleaner

Safety grounded outlet

Insulated rubber gloves

Face shield

Safety glasses

Procedure

1. Obtain permission to use the steam cleaner/pressure washer.
2. Cover all electrical contacts on the machine with the plastic bags and tape.
3. Place the machine at least 15 feet from any building.
4. Plug the cleaner motor into a three-wire grounded outlet. Do not allow the plug to get wet.
5. Attach the water hose to the inlet of the cleaner and the water supply.
6. Put on the safety glasses, face shield, and rubber gloves.
7. Fill the fuel tank on the cleaner according to the manufacturer's instructions.
8. Mix the cleaning compound in the solution tank.
9. Position the cleaner so you can monitor all gauges.
10. Lay out the hose.
11. Hold the nozzle away from anyone and toward the ground.
12. Open the water faucet.
13. Turn on the cleaner or steamer motor.
14. Turn on the fuel valve and start the cleaner.
15. Adjust the fuel valve to create the recommended pressure for the cleaner.
16. When steam comes from the nozzle, adjust the cleaning solution to obtain the desired cleaning action.
17. Start at the top of the machine and hold the nozzle about 8" to 12" from the machine.

18. Hold the nozzle about 2–4" away from tougher places.
19. Steam the whole machine lightly to remove any dirt.
20. Turn off the solution valve.
21. Turn off the fuel valve.
22. Hold the nozzle until cold water comes out.
23. Turn off the motor and the water.
24. Unplug the cleaner at the outlet.
25. Disconnect the hose from the machine and hose down the area.
26. Drain the machine and hoses using compressed air if they are to be exposed to freezing temperatures.
27. Properly store the cleaner.
28. Remove the plastic from the machine.
29. If motorized, start the machine to dry it. Otherwise, hand-dry the equipment.

Instructor's Score or Approval ______________________________

Job Sheet 28-5

Name ____ Date _______________

Preparing New Metal for Painting

Objective

Upon completing this activity, you will have demonstrated the ability to prepare new metal for painting.

Tools and Materials

Job Sheet 28-5

Metal to be painted

Mineral spirits

Rags

Commercial solvent

Emery cloth

Chipping hammer

Wire brush

Safety glasses

Procedure

1. Obtain permission to start the procedure.
2. Put on the safety glasses.
3. Chip and brush any welds.
4. Sand to remove any dirt or rust.
5. Clean with mineral spirits to remove grease or oil.
6. Clean with commercial solvent.
7. Clean the area and put away the materials.

Instructor's Score or Approval ____

Job Sheet 28-6

Name ______________________________________ Date ______________

Preparing Previously Painted Metal

Objective

Upon completing this activity, you will have demonstrated the ability to prepare painted metal for painting.

Tools and Materials

Job Sheet 28-6

Metal to be repainted

Mineral spirits

Power sander

Sandpaper (200 to 600 grit)

Water and pan

Wire brush

Safety glasses

Procedure

1. Obtain permission to start the procedure.
2. Put on the safety glasses.
3. Steam or pressure clean the metal, if needed.
4. If not steamed, clean the metal with scrapers and solvents.
5. Power sand where required.
6. Feather minor chips by water sanding with very fine grit (200) until no edges can be felt.
7. Sand with 400–600 paper until no sanding marks are visible.
8. Clean thoroughly and dry to prevent rust.
9. Clean the area and put away the materials.

Instructor's Score or Approval ____

Job Sheet 28-7

Name ______________________________ Date ____________

Masking for Spray Painting

Objective

Upon completing this activity, you will have demonstrated the ability to mask objects for spray painting.

Tools and Materials

Job Sheet 28-7

Object to be masked

Masking tape and newspapers

Grease

Utility knife

Procedure

1. Use grease to cover any small areas like grease fittings.
2. Use newspaper and masking tape to cover glass and gauges.
3. Mask any areas not being painted by cutting a piece of newspaper to ¼" less than the size of the area, and then tape the edges.
4. Use the utility knife to trim overhanging tape or paper.
5. Mask and tape any areas to be painted a second color.
6. Clean the area and put away the materials.

Instructor's Score or Approval ______________________________

Job Sheet 28-8

Name ______________________________ Date ____________

Estimating Paint Jobs

Objective

Upon completing this activity, you will have demonstrated the ability to estimate the materials needed for a paint job.

Tools and Materials

Job Sheet 28-8

Pen or pencil

Paper

Procedure

Read the following methods for calculating materials. If you do not understand any of the steps, ask your instructor for more information.

A. For buildings:
   1. Determine the square footage to be painted.
   2. Divide the square footage by the coverage of the paint in square feet.
   3. Round up the amount to the next quart.
   4. If this estimate results in three quarts, round to a gallon.
   5. Allow for rougher spots or needed second coats.
   6. If oil-based paint will be used, order thinner or brush cleaner.
   7. Complete the list of paint materials to be ordered.

B. For machinery:

   Remember: A tractor takes a gallon of the main color and a quart of the trim paint.
   1. Compare the size and area of the machine to a small tractor.
   2. Use the same rounding of materials as with the previous exercise.
   3. Order thinner in proportion to the amount of paint.
   4. Order less-expensive materials for cleaning the spray equipment.
   5. Complete your list of materials.

Instructor's Score or Approval ______________________________

UNIT 29

Selecting and Applying Painting Materials

In the last unit, you learned how to prepare surfaces for paint. In this unit, we are concerned with selecting and applying the paint that will beautify or protect wood or metal surfaces.

Class Activity 29-1: Understanding Painting Materials

Class Activity 29-2: Identifying Painting Materials

Shop Activity 29-3: Using Brushes

Shop Activity 29-4: Using Rollers

Shop Activity 29-5: Using Aerosols

Shop Activity 29-6: Using Spray Guns

Shop Activity 29-7: Cleaning Spray Equipment

Job Sheet 29-1

Name ______________________________ Date ____________

Understanding Painting Materials

Objective

Upon completing this activity, you will have demonstrated an understanding of painting materials.

Tools and Materials

Job Sheet 29-1

Pen or pencil

Procedure

Answer the following questions:

1. What are the two main reasons for painting?

____

____

____

2. What is a paint formulation?

____

____

____

3. What is pigment?

____

____

____

4. What differentiates high-quality pigments from low-quality pigments?

____

____

____

5. What are primers or undercoats?

____

____

____

____

6. What do we mean by compatibility of paints?

____

____

7. How can a paint be rust-resistant?

____

____

____

8. Name the parts of the spray gun in the following picture:

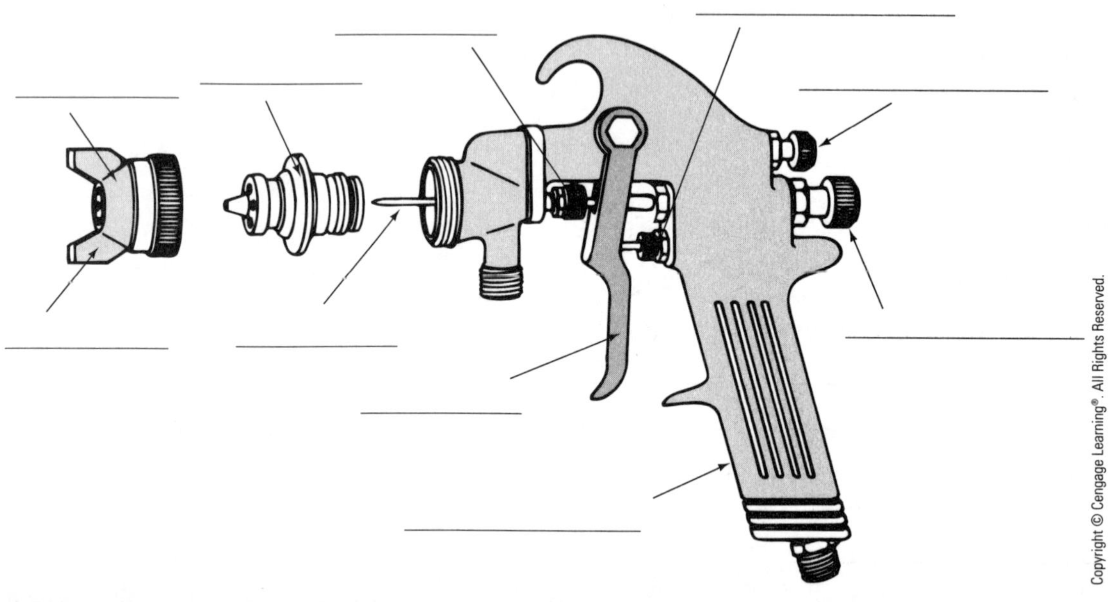

Figure 29-1 Parts of a Spray Gun

Instructor's Score or Approval ______________________________

Job Sheet 29-2

Name ____ Date _______________

Identifying Painting Materials

Objective

Upon completing this activity, you will have demonstrated the ability to identify painting materials.

Tools and Materials

Job Sheet 29-2

Pen or pencil

Procedure

Match the items in Column I with those in Column II.

	Column I	*Column II*
____	1. Alkyd	a. Without shine
____	2. Latex	b. Special adhesive and wear-resistant qualities
____	3. Thinner	c. Material that carries the color in painting
____	4. Epoxy	d. Gives color to paint
____	5. Interior	e. Thins paint
____	6. Exterior	f. Has a light shine
____	7. Pigment	g. Shiny
____	8. Hiding power	h. Prepares surface for a high-quality topcoat
____	9. Gloss	i. Suitable for inside use
____	10. Semigloss	j. Ability to cover previous colors
____	11. Flat	k. Formulated for weather exposure
____	12. Primer	l. Indicates water-based paint
____	13. Vehicle	m. Oil-based

Instructor's Score or Approval ____

Job Sheet 29-3

Name ______________________________________ Date ______________

Using Brushes

Objective

Upon completing this activity, you will have demonstrated the ability to use paintbrushes.

Tools and Materials

Job Sheet 29-3

Paintbrushes

Brush cleaner

Drop cloths

Paint stirrer

Masking tape

Newspaper

Procedure

1. Prepare the surface to be painted.
2. Mask any parts not to be painted.
3. Put drop cloths below the object to be painted.
4. Use masking tape where the paint is to stop.
5. Mix the paint thoroughly.
6. For oil-based paints, use a natural bristle brush; for latex paints, use an artificial bristle brush.
7. Select the brush to fit the job.
8. Dip the bristles one-third of the way into the paint and touch the side to remove excess paint.
9. Touch the brush several places on the area to be painted to deposit the paint.
10. Using the flat side of the brush, smooth the paint, ending the strokes in the painted area.
11. Stir the paint frequently.
12. Paint adjacent areas before the paint dries so the paint coat is uniform.
13. When painting is interrupted, wrap the brush in a plastic bag to prevent the paint hardening in the brush.
14. At the end of the day, clean latex brushes with soap and water and wrap them in paper towels to dry.
15. Oil-based paintbrushes may be suspended in paint thinner or linseed oil overnight.
16. When resuming painting, wipe excess thinner or linseed oil from the brushes with rags.
17. At the completion of the job, clean the brush with paint thinner or brush cleaner.
18. Clean the area and put away the materials.

Instructor's Score or Approval ______________________________________

Job Sheet 29-4

Name ______________________________ Date ____________

Using Rollers

Objective

Upon completing this activity, you will have demonstrated the ability to paint with a roller.

Tools and Materials

Job Sheet 29-4

Paint tray

Roller

Aluminum foil or tray liner

Paint

Drop cloths

Object to be painted

Small brush

Procedure

1. Line the tray with the aluminum foil or a tray liner.
2. Mix the paint.
3. Place the drop cloths.
4. Pour paint into the tray.
5. Paint corners and edges with the small brush first.
6. Roll the larger areas using a short nap roller for smooth surfaces or a longer nap roller for rougher surfaces.
7. Finish by rolling toward the painted area.
8. Stir the paint before refilling the pan.
9. If painting is stopped, wrap the roller in an airtight plastic bag.
10. When done with latex paint, clean quality rollers with soap and water.
11. With oil-based paint or low-quality rollers, discard the covers.
12. Clean the roller handle.
13. Remove and discard the pan liner and clean any remaining paint.
14. Clean the small brush and wrap it to dry.
15. Clean the area and put away the materials.

Instructor's Score or Approval ______________________________

Job Sheet 29-5

Name ______________________________ Date ____________

Using Aerosols

Objective

Upon completing this activity, you will have demonstrated the ability to use aerosol paints.

Tools and Materials

Job Sheet 29-5

Object to be painted

Aerosol paint

Cardboard

Drop cloth

Masking tape

Safety glasses

Respirator

Procedure

1. Prepare the object to be painted by sanding or degreasing as needed.
2. Mask as needed.
3. Choose a well-ventilated area for painting.
4. Shake the aerosol can for several minutes.
5. Put on the safety glasses and the respirator.
6. Practice on a piece of cardboard to see the spray pattern.
7. Spray in rapid uniform strokes, starting in the air and moving across the object.
8. Coat the surface with a mist and, when tacky, build up the thickness needed.
9. When finished, invert the can and spray until no paint emerges.
10. If applying more than one coat, follow the directions on the can.
11. Clean the area and put away the materials.

Instructor's Score or Approval ______________________________

Job Sheet 29-6

Name ______________________________ Date ____________

Using Spray Guns

Objective

Upon completing this activity, you will have demonstrated the ability to use a spray gun in painting.

Tools and Materials

Job Sheet 29-6

Spray equipment

Paint

Thinner

Paint strainer

Rags

Object to be painted

Safety glasses

Paint respirator

Procedure

1. Prepare and mask the object to be painted.
2. Mix the paint thoroughly.
3. Test the thinner with a small amount of paint to be sure they are compatible.
4. Thin the paint to the recommended viscosity.
5. Strain the paint into the cup.
6. Put on the safety glasses and the paint respirator.
7. Turn on the air and set the regulator to the recommended pressure for the gun.
8. Adjust the spray pattern using a cardboard target.
9. Use uniform strokes 6" to 10" from, and parallel to, the surface to obtain a uniform covering without dusting or sagging.
10. Paint the object using passes that overlap the previous pass by 50 percent.
11. Clean the area and put away all the materials but the spray gun. You will use the procedure listed on Job Sheet 29-7 to clean the spray gun.

Instructor's Score or Approval ______________________________

Job Sheet 29-7

Name ______________________________ Date ____________

Cleaning Spray Equipment

Objective

Upon completing this activity, you will have demonstrated the ability to clean paint-spraying equipment.

Tools and Materials

Job Sheet 29-7

Spray equipment used in the previous activity

Paint thinner

Rags

Cardboard

Procedure

1. Pour all unused paint back into the can and reseal it.
2. Wipe most of the paint from the gun.
3. Pour 1" of paint thinner into the gun.
4. Shake it vigorously.
5. Spray the paint thinner onto the cardboard.
6. Wipe all visible surfaces with paint thinner on a cloth.
7. Repeat Steps 3 through 6 until no traces of paint appear.
8. Remove the air and paint nozzle.
9. Use a cloth and thinner to wipe away any traces of paint.
10. Dry all parts and reassemble the gun.
11. Clean the rest of the equipment with solvent and cloth.
12. Discard cleaning rags and cans in an approved metal container.
13. Clean the area and put away the materials.

Instructor's Score or Approval ______________________________

SECTION 10

Power Mechanics

Unit 30

Fundamentals of Small Engines

Unit 31

Small Engine Maintenance and Repair

Unit 32

Diesel Engines and Tractor Maintenance

UNIT 30

Fundamentals of Small Engines

Small engines are a common source of power in agricultural mechanics. Small engines also are used for many applications by homeowners in urban areas. Knowledge of small engine fundamentals is the subject of this unit.

Class Activity 30-1: Understanding Small Engine Fundamentals

Class Activity 30-2: Identifying Safety Precautions with Small Engines

Job Sheet 30-1

Name ______________________________________ Date _______________

Understanding Small Engine Fundamentals

Objective

Upon completing this activity, you will have demonstrated an understanding of small engine fundamentals.

Tools and Materials

Job Sheet 30-1

Pen or pencil

Procedure

Answer the following questions:

1. What is an external combustion engine?

____

____

____

2. What is an internal combustion engine?

____

____

____

3. Why was the development of the internal combustion engine so important to agriculture?

____

____

4. Name the four strokes of the four-cycle engine and describe what is happening in each.

   First stroke: Name

   ____

   ____

   Second stroke: Name

   ____

   ____

   Third stroke: Name

   ____

   ____

   Fourth stroke: Name

   ____

   ____

5. What are the major differences between a four-cycle engine and a two-cycle engine?

6. What are the advantages of the two-cycle engine?

7. What is compression?

8. How is compression created in the cylinder?

9. What are the functions of the valves?

10. Why must valves fit tightly?

11. Name the parts of the fuel system.

12. Describe the venturi.

13. Name the parts of the ignition system.

14. What is a governor?

15. Why is a cooling system needed?

16. What are the four functions of oil in a small engine?

a. ______

b. ______

c. ______

d. ______

Instructor's Score or Approval ______

Job Sheet 30-2

Name ______________________________________ Date ______________

Identifying Safety Precautions with Small Engines

Objective

Upon completing this activity, you will have demonstrated the ability to identify safety precautions with small engines.

Tools and Materials

Job Sheet 30-2

Pen or pencil

Procedure

Match the items in Column I with those in Column II.

Column I	*Column II*
____ 1. Never start an engine	a. overload the engine.
____ 2. Avoid smoking	b. handle gasoline outdoors.
____ 3. Always wear	c. as a cleaner.
____ 4. Keep hands and feet	d. around gasoline or oil.
____ 5. Stop the engine and cool	e. above recommended speeds.
____ 6. Avoid spilling	f. away from moving parts.
____ 7. Whenever possible	g. before making adjustments.
____ 8. Disconnect the spark plug	h. gasoline on hot parts.
____ 9. Do not operate engines	i. in an unventilated area.
____ 10. Do not	j. before refueling.
____ 11. Never use gasoline	k. safety glasses, hearing protection, and leather shoes.

Instructor's Score or Approval ______________________________________

UNIT 31 Small Engine Maintenance and Repair

Small engines provide low-horsepower portable power units for many agricultural and recreational uses. This unit is designed to help you learn to maintain and repair small engine units.

Class Activity 31-1: Understanding Small Engine Maintenance and Repair

Class Activity 31-2: Identifying Safety Precautions

Shop Activity 31-3: Servicing Foam Air Cleaners

Shop Activity 31-4: Servicing Dry Element Air Cleaners

Shop Activity 31-5: Servicing Dual Element Air Cleaners

Shop Activity 31-6: Servicing Oil Bath Air Cleaners

Shop Activity 31-7: Adjusting the Carburetor

Shop Activity 31-8: Changing Crankcase Oil

Shop Activity 31-9: Troubleshooting

Shop Activity 31-10: Servicing Starter Mechanisms

Shop Activity 31-11: Preparing Small Engines for Storage

Shop Activity 31-12: Deciding the Big Question: Repair or Replace?

Job Sheet 31-1

Name ____ Date _______________

Understanding Small Engine Maintenance and Repair

Objective

Upon completing this activity, you will have demonstrated an understanding of small engine maintenance and repair.

Tools and Materials

Job Sheet 31-1

Pen or pencil

Procedure

Answer the following questions:

1. List the basic tools required for small engine maintenance.

____

____

____

____

____

2. ________ With proper attention to maintenance, an engine will usually provide years of service. True or false?
3. ________ Air cleaners are important because they prevent air pollution. True or false?
4. ________ Air cleaners need only be serviced at the 100-hour interval. True or false?
5. ________ All air cleaners are serviced the same way. True or false?
6. ________ Any damaged air cleaner element should be replaced. True or false?
7. ________ Clean fuel is essential to proper operation. True or false?
8. ________ Gasoline that sits for a period of time may lose some of its value. True or false?
9. ________ A partially sheared flywheel key can cause starting problems. True or false?
10. ________ A valve face can be ground to one quarter of its margin. True or false?
11. ________ Starting mechanisms require no maintenance. True or false?
12. ________ Plastigage is used to measure compression. True or false?
13. ________ Carbon deposits can be caused by blowby. True or false?
14. ________ Major parts may cost more to replace than replacing the engine. True or false?
15. ________ Small engine troubleshooting skills are useful to many people. True or false?

Instructor's Score or Approval ____

Job Sheet 31-2

Name ____ Date _______________

Identifying Safety Precautions

Objective

Upon completing this activity, you will have demonstrated the ability to identify maintenance safety precautions.

Tools and Materials

Job Sheet 31-2

Pen or pencil

Procedure

Match the items in Column I with those in Column II.

Column I		*Column II*
____	1. Work only in	a. in the work area.
____	2. An approved fire extinguisher	b. in an approved fireproof container.
____	3. Drain fuels	c. all oil or fuel spills.
____	4. Store fuels	d. found in the engine manual.
____	5. Goggles	e. should be kept in the work area.
____	6. Wipe immediately	f. flammable
____	7. Avoid open flame or sparks	g. only in approved containers.
____	8. Observe all safety precautions	h. a well-ventilated area.
____	9. Place rags with fuel on them	i. should be worn at all times
____	10. Keep ________ materials out of the work area	j. outdoors.

Instructor's Score or Approval ____

Job Sheet 31-3

Name ____ Date _______________

Servicing Foam Air Cleaners

Objective

Upon completing this activity, you will have demonstrated the ability to clean foam air cleaners.

Tools and Materials

Job Sheet 31-3

Small engine with a foam air cleaner

Kerosene

Rags

Motor oil

Safety glasses

Procedure

1. Put on safety glasses.
2. Remove the wing nut and cover.
3. Replace the foam filter if it crumbles or is torn.
4. Wash the foam filter in the kerosene.
5. Squeeze out the excess liquid.
6. Dry the foam filter by squeezing it with the cloth.
7. Soak the foam filter with the recommended motor oil.
8. Squeeze out the excess oil.
9. Clean the metal parts and cover them with the solvent.
10. Reassemble the air cleaner.
11. Clean the area and put away the materials. Remember to put used rags in a fireproof container.

Instructor's Score or Approval ____

Job Sheet 31-4

Name ____ Date _______________

Servicing Dry Element Air Cleaners

Objective

Upon completing this activity, you will have demonstrated the ability to service a dry element air cleaner.

Tools and Materials

Job Sheet 31-4

Small engine with a dry element air cleaner

Kerosene

Rags

Safety glasses

Procedure

1. Put on the safety glasses.
2. Remove the wing nut and cover.
3. If the recommended hours have elapsed, replace the cartridge with a new one.
4. If the element is clogged or torn, replace it.
5. Tap the element on a hard surface to dislodge dirt.
6. Clean the area and the cover with the solvent.
7. Reassemble the air cleaner.
8. Clean the area and put away the materials. Dispose of used rags properly.

Instructor's Score or Approval ____

Job Sheet 31-5

Name ______________________________ Date ____________

Servicing Dual Element Air Cleaners

Objective

Upon completing this activity, you will have demonstrated the ability to service a dual element air cleaner.

Tools and Materials

Job Sheet 31-5

Small engine with dual element air filter

Oil

Kerosene

Rags

Safety glasses

Procedure

1. Put on the safety glasses.
2. Remove the wing nut and cover.
3. Remove the foam precleaner and the dry element.
4. Clean the filters as directed in Job Sheets 31-3 and 31-4.
5. Reassemble the air cleaner.
6. Clean the area and put away the materials. Dispose of used rags properly.

Instructor's Score or Approval ______________________________

Job Sheet 31-6

Name ____ Date ______________

Servicing Oil Bath Air Cleaners

Objective

Upon completing this activity, you will have demonstrated the ability to service an oil bath air cleaner.

Tools and Materials

Job Sheet 31-6

Engine with an oil bath air cleaner

Oil recommended for the engine

Kerosene

Rags

Safety glasses

Procedure

1. Put on the safety glasses.
2. Remove the wing nut and cover.
3. Remove the core.
4. Remove the cleaner body and dump the dirty oil (remember to dispose of it properly).
5. Clean the body with kerosene.
6. Reinstall the body and fill with oil to the mark.
7. Rinse the wire element with the solvent.
8. Shake the excess solvent from the element.
9. Clean and reinstall the cover and wing nut.
10. Clean the area and put away the materials. Dispose of used rags properly.

Instructor's Score or Approval ____

Job Sheet 31-7

Name ______________________________ Date ______________

Adjusting the Carburetor

Objective

Upon completing this activity, you will have demonstrated the ability to adjust a carburetor.

Tools and Materials

Job Sheet 31-7

Small engine with a carburetor

Screwdriver

Tachometer

Safety glasses

Procedure

1. Put on the safety glasses.
2. Start the engine and warm it to operating temperature.
3. Adjust the idle speed adjustment until the engine idles at the recommended speed, as shown on the tachometer.
4. Slowly turn the idle mixture screw clockwise until the engine slows down or labors.
5. Counting the number of turns, slowly turn the screw counterclockwise until the engine runs smoothly and then slows down.
6. Slowly turn the screw clockwise half the number of turns used in Step 5. The engine should now be running more smoothly.
7. Recheck and adjust the idle speed.
8. Now set the engine at full throttle.
9. Repeating Steps 4 through 6, turn the fuel-air mixture screw until the engine runs smoothly.
10. Put the engine under load and check its performance.
11. If the engine does not accelerate or pull well, slightly turn the fuel mixture screw counterclockwise to enrich the mixture.
12. Clean the area and put away the materials.

Instructor's Score or Approval ______________________________

Job Sheet 31-8

Name ______________________________ Date ____________

Changing Crankcase Oil

Objective

Upon completing this activity, you will have demonstrated the ability to change the crankcase oil.

Tools and Materials

Job Sheet 31-8

Small engine needing an oil change

Recommended oil

Adjustable wrench

Oil drain pan

Funnel

Safety glasses

Procedure

1. Put on the safety glasses.
2. Warm the engine and shut it off. Remove the spark plug wire.
3. Place the drain pan under the drain plug and remove the plug.
4. Allow the oil to completely drain.
5. Replace the drain plug.
6. Fill the crankcase with the recommended amount of clean oil.
7. Clean any spills with the rags.
8. Clean the area and put away the materials. Dispose of the used oil and rags properly.

Instructor's Score or Approval ______________________________

Job Sheet 31-9

Name ______________________________ Date ____________

Troubleshooting

Objective

Upon completing this activity, you will have demonstrated the ability to troubleshoot a small engine.

Tools and Materials

Job Sheet 31-9

Small engine that will not start

Small engine toolkit

Safety glasses

Procedure

1. Put on the safety glasses.
2. Check that the plug wire is properly installed and has no breaks.
3. Test for spark; for no spark, check all kill switches and wiring for grounding.
4. If the spark is weak, change the plug.
5. If the spark is good, check for a partially sheared flywheel key.
6. Check the fuel tank for fuel.
7. Check the plug for the odor of gasoline after trying to start the engine.
8. If no odor is present, check the fuel lines for any obstruction or fuel filter faults.
9. If the plug is wet, check for flooding due to the choke setting or a stuck needle valve.
10. If flooded, let the engine sit for 10 minutes and try again.
11. If the engine runs but does not produce power, a major overhaul may be needed. (Due to the many different engines available, the procedure for overhaul needs to be followed as given in the operator's manual for the specific engine.)
12. Clean the area and put away the materials.

Instructor's Score or Approval ______________________________

Job Sheet 31-10

Name ______________________________________ Date ______________

Servicing Starter Mechanisms

Objective

Upon completing this activity, you will have demonstrated the ability to service starting mechanisms.

Tools and Materials

Job Sheet 31-10

Small engine with a starter problem

Repair manual for the engine

Small engine tools

Safety glasses

Procedure

1. Obtain the repair manual for your engine.
2. Put on the safety glasses.
3. Follow the repair manual procedure for the type of starter mechanism on your engine. Disassemble the starter carefully because of the spring and sharp edges.
4. Reattach a new starter rope, cleaning all parts.
5. Reassemble the starter mechanism.
6. Clean the area and put away the materials.

Instructor's Score or Approval ______________________________________

Job Sheet 31-11

Name ______________________________ Date ____________

Preparing Small Engines for Storage

Objective

Upon completing this activity, you will have demonstrated the ability to prepare a small engine for storage for a period of time.

Tools and Procedures

Job Sheet 31-11

Small engine to be prepared for storage

Recommended oil

Rags

Solvent

Safety glasses

Procedure

1. Put on the safety glasses.
2. Drain the fuel tank.
3. Shut off the fuel valve.
4. Run the engine until all the fuel is burned.
5. Drain the crankcase of a four-cycle engine and refill it with the recommended oil.
6. Remove the spark plug (for two-cycle engines, rotate the crankshaft until the piston covers the ports) and add 2–3 tablespoons of clean engine oil.
7. Reinstall the spark plug.
8. Rotate the crankshaft several times to distribute the oil.
9. Drain, clean, and reinstall the fuel filter bowl.
10. Remove the shrouds and guards so the engine can be cleaned thoroughly.
11. Clean the exterior.
12. Remove any debris from the fins.
13. Check all linkages for free movement.
14. Loosen all belts.
15. Clean and lubricate all chains.
16. Lubricate all fittings.
17. Reinstall all shrouds and guards.
18. Store the engine off the ground.
19. If outdoors, cover the engine to protect it.
20. Clean the area and put away the materials.

Instructor's Score or Approval ______________________________

Job Sheet 31-12

Name ______________________________ Date ____________

Deciding the Big Question: Repair or Replace?

Objective

Upon completing this exercise, you will have demonstrated the ability to decide whether to overhaul or replace a small engine.

Tools and Materials

Job Sheet 31-12

Small engine in need of repair

Repair manual for the engine

Small engine tools, including specialized items

Rags

Kerosene

Torque wrench

Plastigage

Micrometers, inside and outside

Safety glasses

Procedure

1. Obtain the manual for your engine.
2. Put on the safety glasses.
3. Follow the disassembly procedures, including using the ridge reamer to prevent breaking the piston rings upon removing the piston.
4. Use the piston ring expander to remove the piston rings.
5. Using the inside micrometer, measure the cylinder bore parallel and at right angles to the direction of the wrist pin at several different levels within the piston's travel range.
6. Compare the readings with the manufacturer's tolerances.
7. Check the cylinder walls for scoring and wear.
8. Check the crankshaft bearing for scoring and out-of-round with the micrometer.
9. Compare the readings to the tolerances.
10. Cut a piece of Plastigage, put it in the bearing cap, and torque the cap to the specified inch or foot pounds.
11. Measure the flattened Plastigage.
12. Compare the clearance to tolerances.
13. Place the solid piston ring in the cylinder and measure the ring gap clearance with a feeler gauge.
14. Compare the reading to the tolerances.
15. Check the cost of any replacement parts needed.
16. Compare the cost of needed parts with the expected wear and the cost of a replacement.

17. Share your comparison measurements with your instructor and agree on a plan of action.
18. If the cost seems reasonable in comparison to the engine's expected life, order the parts and reassemble the engine.
19. Clean the area and put away the materials.

Instructor's Score or Approval ____

UNIT 32

Diesel Engines and Tractor Maintenance

Diesel engines are commonly used in agricultural mechanics. Knowledge of how they work and the proper maintenance procedures are key in keeping your equipment running. This is the subject of this unit.

Class Activity 32-1: Understanding Diesel Engine Fundamentals

Class Activity 32-2: Understanding Engine Maintenance

Shop Activity 32-3: Checking the Cooling System

Job Sheet 32-1

Name ________________________________ Date ______________

Understanding Diesel Engine Fundamentals

Objective

Upon completing this activity, you will have demonstrated an understanding of diesel engine fundamentals.

Tools and Materials

Job Sheet 32-1

Pen or pencil

Procedure

Answer the following questions:

1. What is the diesel engine?

____

____

____

2. What are the differences between the use of early diesel engines and how they are used today?

____

____

____

3. What are the disadvantages of using a diesel engine?

____

____

____

4. What are the advantages of using a diesel engine?

____

____

____

5. What is the difference between gasoline and diesel fuel?

____

____

____

6. What is the difference between the strokes of a four-cycle gasoline engine and the diesel engine?

____

____

____

7. What is a compression ratio?

____

____

____

8. Why must the components of the diesel engine be heavier and more durable than those of the gasoline engine?

____

____

____

9. How does the diesel engine produce more power than the gasoline engine?

____

____

____

10. Explain where the diesel engine's "knock" comes from.

____

____

____

Instructor's Score or Approval ______________________________

Job Sheet 32-2

Name ______________________________ Date ____________

Understanding Engine Maintenance

Objective

Upon completing this activity, you will have demonstrated an understanding of diesel engine maintenance.

Tools and Materials

Job Sheet 32-2

Pen or pencil

Procedure

Answer the following questions:

1. What is the most vital component of the engine system?

2. What is the function of the oil in the engine system?

3. Why must the oil be changed at regular intervals?

4. How do you know the proper grade, viscosity, and intervals for changing the oil in your engine?

5. What do SAE ratings tell us about an oil?

6. What is the difference between an SE and a CA oil classification?

7. What are the differences between different oil weights, and when would you need to use them?

8. Name four oil additives and discuss what each one does.

____

____

____

____

____

9. What is the function of the oil filter?

____

____

____

10. When does the oil filter need to be replaced?

____

____

____

11. How do you maintain the air filter?

____

____

____

Instructor's Score or Approval ______________________________

Job Sheet 32-3

Name ______________________________ Date ____________

Checking the Cooling System

Objective

Upon completing this activity, you will have demonstrated the ability to check the cooling system on a tractor.

Tools and Materials

Job Sheet 32-3

Small tractor engine

Hydrometer

Manufacturer's operator manual

Garden hose

Safety glasses

Procedure

1. Obtain permission to start the procedure.
2. Put on the safety glasses.
3. Remove the radiator cap and check the coolant level. **Caution!!!** Make sure the engine is cool before removing the radiator cap.
4. Check the gasket or radiator cap. Replace if needed.
5. Use the hydrometer to see if the coolant needs to be changed.
6. If liquid needs to be added or replaced, follow the manufacturer's manual to mix the proper amounts of coolant and water.
7. Put the cap back on the radiator.
8. Use the hose to wash the radiator and remove all debris.
9. Clean the area and put away the materials.

Instructor's Score or Approval ______________________________

SECTION 11

Electricity and Electronics

UNIT 33

Electrical Principles and Wiring Materials

Electricity is a major source of energy for many operations in agriculture, in the home, and in industry. An understanding of electrical principles will be useful to you in any occupation.

Class Activity 33-1: Understanding Electrical Principles

Shop Activity 33-2: Electric Motor Maintenance

Class Activity 33-3: Identifying Electrical Safety Precautions

Class Activity 33-4: Identifying Standard Electrical Symbols

Job Sheet 33-1

Name ______________________________ Date ____________

Understanding Electrical Principles

Objective

Upon completing this activity, you will have demonstrated an understanding of electrical principles.

Tools and Materials

Job Sheet 33-1

Pen or pencil

Procedure

Answer the following questions:

1. __________ is the major power source for stationary equipment in homes, on farms, and in businesses.
2. Electricity can produce heat, light, magnetism, and chemical changes. True or false?
3. What is an electrical insulator?

   ____

   ____

4. State Ohm's Law.

   ____

   ____

   ____

5. What devices are used to measure volts, amperes, and watts?

   ____

   ____

6. Define magnetism.

   ____

   ____

7. What is a permanent magnet?

   ____

   ____

8. What is the relationship between magnets and electricity?

   ____

   ____

   ____

9. What are magnetic poles?

   ____

   ____

10. Which magnetic poles attract each other?

   ____

11. Which magnetic poles repel each other?

____

12. Magnetic lines of force are called ____________________.
13. What are electromagnets?

____

____

14. Why are electromagnets important?

____

____

15. What is an armature?

____

____

16. What is a generator?

____

____

17. What is an alternator?

____

____

18. What periodic maintenance does an electric motor need?

____

____

19. What is the best source of information about electric motor maintenance?

____

____

20. What is a branch circuit?

____

____

21. How are branch circuits protected?

____

____

22. What is nonmetallic sheathed cable?

____

____

23. What is armored cable?

____

____

24. What does the term *wire gauge* mean?

____

____

25. How can you identify the gauge of a wire?

____

____

26. What are the relationships among wire size, length of run, and voltage drop?

____

27. Cite the recommended use for each of the following wire types.

Type	Recommended Use
T	______________________
TW	______________________
THHN	______________________
THW or THWN	______________________
XHHW	______________________
UF	______________________

Instructor's Score or Approval ______________________

Job Sheet 33-2

Name ____ Date ______________

Electric Motor Maintenance

Objective

Upon completing this activity, you will have demonstrated the ability to do electric motor maintenance.

Tools and Materials

Job Sheet 33-2

Electric motor

Compressed air source

Oil

Solvent

Owner's manual for the motor

Rags

Safety glasses

Procedure

1. Obtain permission to start the procedure.
2. Put on the safety glasses.
3. Using compressed air, blow out dust and debris from the exterior parts of the motor.
4. Using the rag and solvent, clean any remaining dirt or grease from the motor case.
5. Oil the motor according to the owner's manual.
6. Clean the area and put away the materials.

Instructor's Score or Approval ____

Job Sheet 33-3

Name ______________________________________ Date ______________

Identifying Electrical Safety Precautions

Objective

Upon completing this activity, you will have demonstrated the ability to identify safety precautions with electricity.

Tools and Materials

Job Sheet 33-3

Pen or pencil

Procedure

Match the items in Column I with those in Column II.

	Column I	*Column II*
____	1. Safety devices	a. in the kitchen, bath, and anywhere moisture conditions exist.
____	2. Do not touch electrical devices	b. worn electrical cords.
____	3. Do not remove	c. under carpets.
____	4. Use ground fault circuit interrupters	d. National Electrical Code®.
____	5. Discontinue use of	e. with wet hands.
____	6. Do not place extension cords	f. must not be disabled.
____	7. Install all wiring according to the	g. the third prong on appliance cords.
____	8. When a fuse blows or circuit breaker opens	h. for installation of any electrical equipment.
____	9. Use only double insulated or three-wire grounded cord	i. unattended.
____	10. Do not install	j. determine the cause before restoring power.
____	11. Do not leave heat-producing appliances	k. to power portable tools.
____	12. Keep all heaters and lamps away from	l. higher amperage fuses or circuit breakers than the rating of the circuits they protect.
____	13. Keep metal cases of electrical appliances	m. combustible materials.
____	14. Don't use cracked or broken	n. to reduce shock hazard and prevent rust.
____	15. Follow the manufacturer's recommendations	o. switches or outlets.
____	16. Keep appliances dry	p. grounded at all times.

Instructor's Score or Approval ______________________________________

Job Sheet 33-4

Name ______________________________ Date ____________

Identifying Standard Electrical Symbols

Objective

Upon completing this activity, you will have demonstrated the ability to identify the standard electrical symbols.

Tools and Materials

Job Sheet 33-4

Pen or pencil

Procedure

Complete the worksheet on the next page.

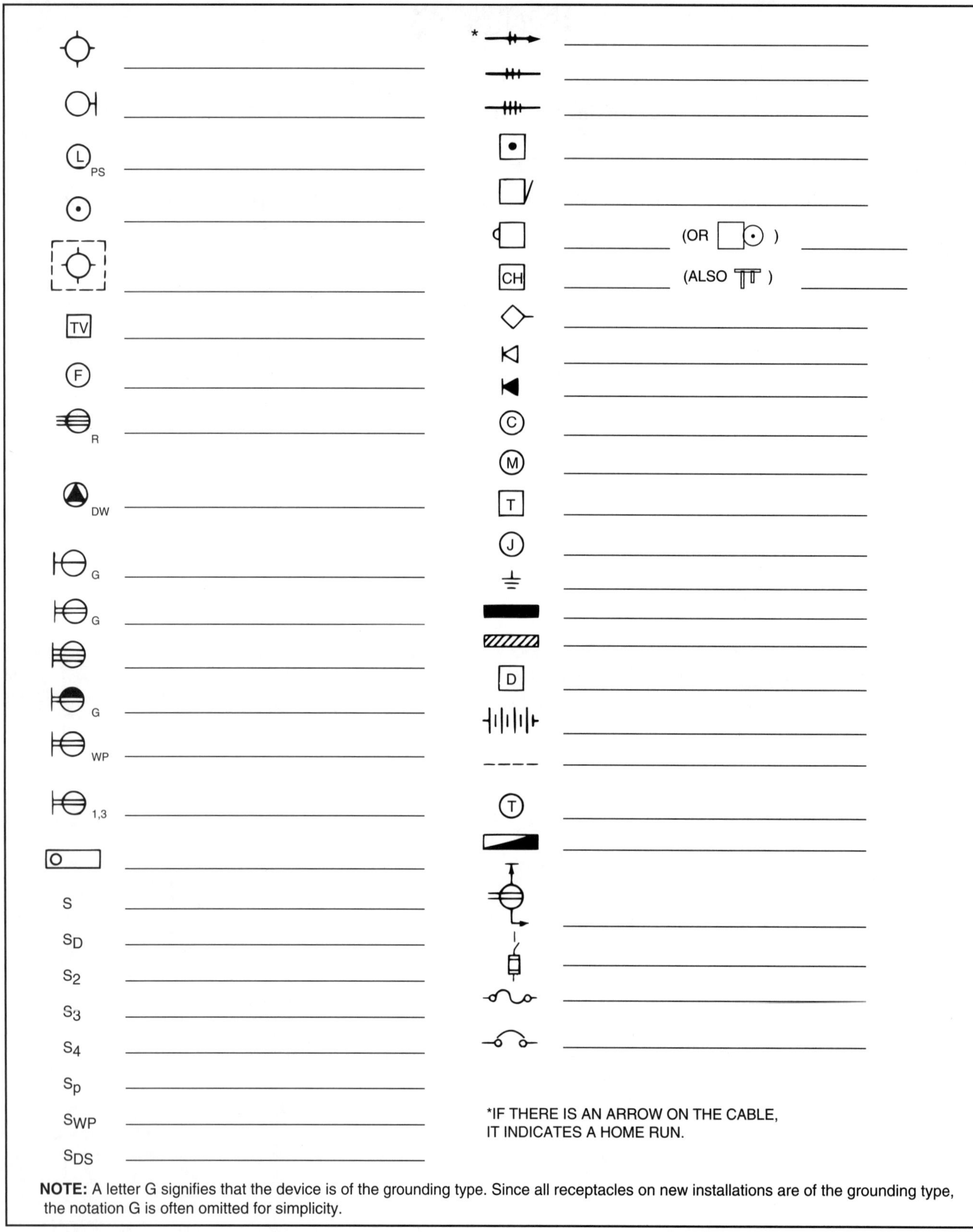

Figure 33-1 Electrical Symbols

Instructor's Score or Approval ______________________

UNIT 34

Installing Branch Circuits

All electrical distribution circuits are called branch circuits when they originate from the main entrance panel. The ability to properly wire a branch circuit will be useful in maintaining a home wiring system.

Class Activity 34-1: Understanding Branch Circuits

Shop Activity 34-2: Wiring a Single-Pole Switch and Lamp Holder with the Power Source at the Lamp Holder

Shop Activity 34-3: Wiring a Single-Pole Switch and Lamp Holder with the Power Source at the Switch

Shop Activity 34-4: Wiring a Duplex Outlet

Shop Activity 34-5: Wiring Three-Way Switches and a Lamp Holder with the Power Source at One Switch and the Lamp Holder between the Switches

Shop Activity 34-6: Wiring Three-Way Switches with the Power Source at the Lamp Holder between the Switches

Shop Activity 34-7: Wiring Three-Way Switches with the Power Source at One Switch and the Lamp Holder beyond the Switches

Shop Activity 34-8: Wiring a Circuit with One or More Four-Way Switches

Job Sheet 34-1

Name ______________________________ Date ____________

Understanding Branch Circuits

Objective

Upon completing this activity, you will have demonstrated an understanding of branch circuits.

Tools and Materials

Job Sheet 34-1

Pen or pencil

Procedure

Answer the following questions:

1. What is the National Electrical Code®? Why should all wiring meet this code?
   ____
   ____
   ____
2. What are wiring boxes?
   ____
3. Why are wiring boxes required?
   ____
4. List the wiring requirements for electrical box installation.
   ____
   ____
   ____
   ____
   ____
5. What is an electrical switch?
   ____
6. How should a white wire be marked in a switch circuit?
   ____
7. Why should a continuity tester be used to test a circuit before it is turned on?
   ____
   ____
   ____

Instructor's Score or Approval ______________________________

Job Sheet 34-2

Name ______________________________ Date ____________

Wiring a Single-Pole Switch and Lamp Holder with the Power Source at the Lamp Holder

Objective

Upon completing this activity, you will have demonstrated the ability to wire a single-pole switch and lamp holder with the power source at the lamp holder.

Tools and Materials

Job Sheet 34-2

Flat-head and Phillips-head screwdrivers

Single-pole switch

Hammer

Switch cover

Needle-nose pliers

Wire strippers

Lamp holder

Pocketknife or jacket ripper

Linesman pliers

Wall box

Wire nuts

Cable staples

Metal ceiling box

Grounding clamps

Volt-ohm meter

Black electrician's tape

14-2 cable with ground

Procedure

1. Install the wall box and the ceiling box.
2. Install the 14-2 cable with ground from the entrance panel to the ceiling box and from the ceiling box to the wall box. The cable should extend 6–8 inches from each box.
3. Support and fasten wires according to the National Electrical Code®.
4. Slit the cable jacket on the cables extending out of the boxes in order to expose the wires.
5. At the wall box, fasten the bare or green wire to the green screw on the switch. If the box is metal, bundle two 8" bare wires to the ground wire, and attach one wire to the switch and one wire to a grounding clip on the wall box.
6. Strip ⅝" of the white and black wires.

7. With the needle-nose pliers, make a clockwise loop at the end of each wire.
8. Place the wires under the two screws on the single-pole switch and tighten. Mark the white wire with black tape.
9. Fasten the switch to the wall box with the screws provided.
10. At the ceiling box, fasten the two black wires together with a wire nut and tighten.
11. Cut an 8" piece of the bare wire and bundle the three bare wires together. Fasten with a wire nut.
12. Fasten the end of the loose bare wire to the ceiling box with a clip or screw (if the boxes are plastic, no grounding is possible).
13. Wrap a piece of black tape around the white wire coming from the wall box because, in the switch circuit, it becomes a hot wire.
14. Make a clockwise loop at the end of the marked wire and place the loop around the brass screw.
15. Place the white wire around the silver screw on the lamp holder.
16. Fasten the lamp holder to the box with the screws provided.
17. With the switch turned on, test the circuit by touching the white and black leads at the entrance panel with a volt-ohm meter. There should be no reading.
18. Test the circuit with a light bulb installed. There should be a reading.
19. If the boxes are metal, check the boxes and the bare wires for continuity.
20. Install the switch plate cover.
21. Wire the circuit into the circuit breaker.
22. Clean the area and put away the materials.

Instructor's Score or Approval ______________________________

Job Sheet 34-3

Name ______________________________ Date __________

Wiring a Single-Pole Switch and Lamp Holder with the Power Source at the Switch

Objective

Upon completing this activity, you will have demonstrated the ability to wire a single-pole switch with the power source at the switch.

Tools and Materials

Job Sheet 34-3

Flat-head and Phillips-head screwdrivers

Single-pole switch

Plastic wall box

Hammer

Lamp holder

Metal ceiling box

Cable staples

Black electrician's tape

14-2 cable with ground

Linesman pliers

Volt-ohm meter

Switch plate cover

Wire strippers

Needle-nose pliers

Grounding clamps

Wire nuts

Pocketknife or jacket ripper

Procedure

1. Install the boxes.
2. Install the cables.
3. Strip the cable coverings back 6 to 8 inches.
4. Strip ⅝" of insulation off each white and black wire.
5. In the switch box, make clockwise loops on the black wire ends.
6. Cut an 8" piece of bare wire and bundle the three bare wires with a wire nut.
7. Ground the loose bare wire to the grounding screw on the switch.
8. Place the white wires in a wire nut and tighten.
9. Place the loops in the black wires around the two terminals on the single-pole switch.

10. Install the switch in the wall box.
11. Clip the bare wire to the ceiling box.
12. Attach the black wire to the brass screw on the lamp holder.
13. Attach the white wire to the silver screw on the lamp holder.
14. Test the circuit with the switch on.
15. Install a lamp bulb and test the circuit again.
16. Test the box's grounding.
17. Connect the circuit to the circuit breaker.
18. Clean the area and put away the materials.

Instructor's Score or Approval ______________________________

Job Sheet 34-4

Name ______________________________ Date ____________

Wiring a Duplex Outlet

Objective

Upon completing this activity, you will have demonstrated the ability to wire a duplex outlet.

Tools and Materials

- Job Sheet 34-4
- Circuit tester
- Duplex outlet
- Metal wall box
- Needle-nose pliers
- Linesman pliers
- 12-2 cable with ground
- Flat-head and Phillips-head screwdrivers
- Wire nuts
- Grounding clamps
- Cable staples
- Hammer

Procedure

1. Install the wall box.
2. Install the cable from the entrance panel to the wall box.
3. Strip 8" off the end of the cable cover.
4. Strip ⅝" of the white and black insulation off the wires.
5. Cut two 8" pieces of bare wire and connect them with the ground wire using a wire nut.
6. Ground the metal box with one of the bare copper wires.
7. Make a clockwise loop in the loose bare wire and install it under the green screw on the outlet.
8. Make loops and install the black wire under the brass screw and the white wire under the silver screw.
9. Install the duplex outlet with the attached screws.
10. Test the circuit for continuity.
11. Wire the circuit to the circuit breaker.*
12. Install the duplex cover.
13. Clean the area and put away the materials.

*Note: If the circuit is to continue to another outlet, there will be four ground wires in the grounding wire nut. A second black wire and a second white wire will attach to the duplex outlet on the second brass terminal and the second silver terminal.

Instructor's Score or Approval ______________________________

Job Sheet 34-5

Name ______________________________________ Date ______________

Wiring Three-Way Switches and a Lamp Holder with the Power Source at One Switch and the Lamp Holder between the Switches

Objective

Upon completing this activity, you will have demonstrated the ability to wire three-way switches with a lamp holder located between them.

Tools and Materials

Job Sheet 34-5

Two three-way switches

Lamp holder

Two plastic wall boxes

Metal ceiling box

Wire stripper

Linesman pliers

Needle-nose pliers

Phillips-head screwdriver

Flat-head screwdriver

14-2 cable with ground

14-3 cable with ground

Wire nuts

Grounding clamps

Pocketknife or jacket ripper

Cable staples

Hammer

Procedure

1. Install the boxes.
2. Install the 14-2 wire from the entrance panel to the first switch.
3. Install the 14-3 wire between both wall boxes and the ceiling box.
4. Strip 8" of the cable jacket at each box.
5. Strip ⅝" of insulation off the end of each white and black wire.
6. At the power source switch, cut an 8" piece of bare wire and connect the three ground wires with a wire nut.
7. Ground the loose wire to the grounding screw on the switch.
8. Connect the two white wires with a wire nut.
9. Attach the black wire from the power source to the single terminal at one end of the switch.

10. Attach the black wire and the red wire from the three-wire cable to the two terminals on the other end of the switch.
11. Install the switch with the screws provided.
12. At the ceiling box, cut an 8" bare wire and connect it with the other two ground wires using a wire nut.
13. Ground the box with the loose bare wire.
14. Attach the two black wires with a wire nut.
15. Attach the two red wires with a wire nut.
16. Attach the white wire from the power source switch to the silver terminal on the lamp holder.
17. Attach the white wire from the second switch (wrap with black tape) to the brass terminal on the lamp holder.
18. Install the lamp holder with the screws provided.
19. At the second switch, ground the bare wire to the switch.
20. Attach the red and black wires to the two wires at one end of the switch.
21. Wrap black tape on the white wire and attach it to the single screw on the other end of the switch.
22. Install the switch with the screws provided.
23. Test the circuit for continuity with the switches in both positions and with a light bulb installed.
24. Install the switch covers.
25. Wire the circuit to the circuit breaker.
26. Clean the area and put away the materials.

Instructor's Score or Approval ______________________________

Job Sheet 34-6

Name ______________________________ Date ____________

Wiring Three-Way Switches with the Power Source at the Lamp Holder between the Switches

Objective

Upon completing this activity, you will have demonstrated the ability to wire three-way switches with the power source at the lamp holder between the switches.

Tools and Materials

Job Sheet 34-6

Black electrician's tape

Wire strippers

Linesman pliers

Two three-way switches

Pocketknife or jacket ripper

14-3 cable with ground

Flat-head and Phillips-head screwdrivers

Lamp holder

14-2 cable with ground

Cable staples

Two switch covers

Two wall boxes

Hammer

Grounding clamps

Wire nuts

Metal ceiling box

Needle-nose pliers

Procedure

1. Install the boxes.
2. Install the 14-2 cable to the ceiling box from the entrance panel.
3. Install the 14-3 cable from each switch box to the ceiling box.
4. Strip 6–8" of the jacket off the ends of the cables.
5. Strip ⅝" of insulation off the ends of the black and white wires.
6. At the ceiling box, cut an 8" piece of bare wire and connect the three ground wires with a wire nut.
7. Ground the loose bare wire to the box with a grounding clip.
8. Attach the black wire from the power source to the white wire from switch number one, and wrap the white wire with the black tape.
9. Connect the black wires from each switch with a wire nut.

10. Connect the red wires from each switch with a wire nut.
11. Attach the white wire from switch number two to the brass screw on the lamp holder. Mark this wire with black tape.
12. Attach the white wire from the power source to the silver screw on the lamp holder.
13. Install the lamp holder with the screws supplied.
14. At switch number two, ground the bare wire to the box with the grounding clip.*
15. Wrap the white wire with black tape and attach it to the single screw on one end of the switch.
16. Connect the black and red wires to the other two screws.
17. Install the switch with the screws provided.
18. At switch number one, ground the bare wire to the box with a grounding clip.*
19. Attach the white wire to the single screw at one end of the switch after marking it with black tape.
20. Attach the red and black wires to the other two screws at the other end of the switch.
21. Attach the switch to the box with the screws provided.
22. Test the circuit.
23. Wire the circuit to the circuit breaker.
24. Install the switch covers.
25. Clean the area and put away the materials.

*Note: When the switches have a green screw, you will need to add two 8" bare wires with a wire nut to ground both the switch and the box.

Instructor's Score or Approval ____________________________

Job Sheet 34-7

Name ______________________________ Date ____________

Wiring Three-Way Switches with the Power Source at One Switch and the Lamp Holder beyond the Switches

Objective

Upon completing this activity, you will have demonstrated the ability to wire three-way switches with the lamp holder beyond the switches.

Tools and Materials

Job Sheet 34-7

Cable staples

Two three-way switches

Hammer

Lamp holder

Linesman pliers

Two wall boxes

Wire strippers

One ceiling box

Pocketknife or jacket ripper

14-3 cable with ground

14-2 cable with ground

Grounding clamps

Wire nuts

Circuit tester

Flat-head and Phillips-head screwdrivers

Procedure

1. Install the wall and ceiling boxes.
2. Install the 14-3 cable between the two wall boxes.
3. Install the 14-2 cable between the entrance panel and the first wall box and between the second wall box and the ceiling box.
4. Strip 8" of the jacket off the cable ends.
5. Strip ⅝" of insulation off the ends of the black, white, and red wires.
6. At the first wall box, install the three-way switch as in Job Sheet 34-5.*
7. At the second wall box, cut an 8" piece of bare wire and connect it with the other two ground wires with a wire nut.
8. Connect the loose bare wire to the box with a grounding clamp.*

9. Connect the two white wires with a wire nut.
10. Connect the red wire to one of the two screws at one end of the switch, and the black wire to the other screw.
11. Connect the black wire from the lamp holder to the single screw at the other end of the switch.
12. Install the switch with the screws provided.
13. At the ceiling box, ground the bare wire to the box with the grounding clip.
14. Attach the black wire to the brass screw and the white wire to the silver screw.
15. Install the lamp holder.
16. Correct the circuit at the circuit breaker.
17. Test the circuit.
18. Install the switch covers.
19. Clean the area and put away the materials.

*Remember: If the box is a metal box or the switch has a green screw, both need to be grounded.

Instructor's Score or Approval ______________________________

Job Sheet 34-8

Name ____ Date ______________

Wiring a Circuit with One or More Four-Way Switches

Objective

Upon completing this activity, you will have demonstrated the ability to wire a circuit with one or more four-way switches.

Tools and Materials

Job Sheet 34-8

Needle-nose pliers

Two three-way switches

Flat-head and Phillips-head screwdrivers

One four-way switch

Three switch covers

Cable staples

Lamp holder

Three wall boxes

Hammer

14-2 cable with ground

Ceiling box

Wire strippers

14-3 cable with ground

Grounding clamps

Pocketknife or jacket ripper

Linesman pliers

Wire nuts

Procedure

1. Install the three wall boxes in a row, followed by the ceiling box.
2. Install the 14-2 cable from the entrance panel to the first wall box and from the third wall box to the ceiling box.
3. Install the 14-3 cable between the first, second, and third wall boxes.
4. Strip 8" of the jacket off the ends of the cables.
5. Strip ⅝" of insulation off the black, white, and red wires.
6. At the first wall box, install a three-way switch as in Job Sheet 34-7.*
7. At the second wall box, cut 8" of bare wire and connect the three ground wires with a wire nut.

8. Ground the loose bare wire to the box.*
9. Connect the two white wires with a wire nut.
10. Connect the red and black wires from the first switch to the two terminals on one end of the four-way switch.
11. Connect the red and black wires leading to the third box to the other two terminals on the other end of the switch.
12. Install the four-way switch with the screws provided.
13. Wire the three-way switch at the third box as in Job Sheet 34-7.
14. Wire the lamp holder as in Job Sheet 34-7.
15. Test the circuit.
16. Install the switch covers.
17. Connect the circuit at the circuit breaker.
18. Clean the area and put away the materials.

* Remember: If the box is metal or the switch has a green screw, both need to be grounded.

Instructor's Score or Approval ______________________________

UNIT 35

Electronics in Agriculture

More and more of our equipment items have one or more functions controlled electronically. Thus, some understanding of the uses and care of electronic equipment will be useful to you.

Class Activity 35-1: Understanding Electronics

Shop Activity 35-2: Maintaining Electronic Equipment

Job Sheet 35-1

Name ______________________________ Date ____________

Understanding Electronics

Objective

Upon completing this activity, you will have demonstrated an understanding of electronics.

Tools and Materials

Job Sheet 35-1

Pen or pencil

Procedure

Answer the following questions:

1. ____________ is said to be the most pervasive technology of the twentieth century.
2. Name five of our current electronic aids.

    ____

    ____

3. What is the unit of measurement for resistance?

    ____

4. Define conductance.

    ____

    ____

5. What is the unit of measurement for conductance?

    ____

6. How are small resistance modules identified?

    ____

7. What are the purposes of electrical multimeters?

    ____

    ____

8. What is the difference between an analog and a digital meter?

    ____

    ____

9. How is a resistance measurement taken?

    ____

    ____

10. How are amperes in a circuit measured?

    ____

    ____

11. How is the voltage in a circuit measured?

12. Define inductance.

13. What is the unit of measurement for inductance?

14. Define capacitance.

15. Define direct current.

16. Define alternating current.

17. Define the term *cycle*.

18. What is an oscilloscope?

19. What are semiconductors?

20. What are transistors?

21. What are thyristors?

22. What are integrated circuits?

23. Describe an optoelectrical device.

24. How are optoelectrical devices used?

25. What is a power supply?

____

____

26. What is an amplifier?

____

____

27. What is an oscillator?

____

____

28. A digital computer processes information using the _______________ number system.
29. Computer chips are the means of storing, processing, or printing many bits of information. True or false?
30. CDs, DVDs, and flash drives are computer _______________ devices.
31. With the development of _______________ and computer _______________, computers have become smaller and more reliable.
32. Name four common applications of microcomputers.

____

____

33. The _______________ provides the mechanism for many agricultural applications.
34. Name one application of lasers in agriculture.

____

35. How do electronics play a part in the modern planter?

____

____

36. Name three other applications of electronics in agriculture.

____

____

37. What is a global positioning system?

____

____

Instructor's Score or Approval _______________________________

Job Sheet 35-2

Name ______________________________ Date ____________

Maintaining Electronic Equipment

Objective

Upon completing this activity, you will have demonstrated the ability to maintain electronic equipment.

Tools and Materials

Job Sheet 35-2

Pen or pencil

Procedure

Complete the statements in Column I by matching them with the statements in Column II.

	Column I	*Column II*
____	1. Read and follow	a. dust and foreign materials.
____	2. When electronics malfunction,	b. a clean dry cloth.
____	3. Maintain a cool and dry	c. the owner's manual.
____	4. Wipe equipment routinely with	d. call a qualified repair technician.
____	5. Lubricate equipment	e. environment for most electronic equipment.
____	6. Keep the equipment free of	f. as recommended by the manufacturer.

Instructor's Score or Approval ______________________________

UNIT 36

Electric Motors, Drives, and Controls

Because electric motors are compact, economical, efficient, safe, and easy to control, they are widely used in agriculture, homes, and industry. The understanding of their installation and control is a usable skill in agricultural mechanics.

Class Activity 36-1: Understanding Electric Motors, Drives, and Controls

Class Activity 36-2: Electric Motor Selection

Class Activity 36-3: Identifying Electrical Control Switches

Shop Activity 36-4: Maintaining Motors, Drives, and Controls

Job Sheet 36-1

Name ______________________________ Date ____________

Understanding Electric Motors, Drives, and Controls

Objective

Upon completing this activity, you will have demonstrated an understanding of electric motors, drives, and controls.

Tools and Materials

Job Sheet 36-1

Pen or pencil

Procedure

Answer the following questions:

1. The textbook suggests that electric motors are the ____________ of modern technology.
2. List 10 reasons electric motors are commonly chosen over other types of power supplies.

____

____

____

____

____

____

____

____

____

____

3. Name the factors to be considered when selecting an electric motor.

____

____

____

4. What information is found on an electric motor nameplate?

____

____

____

5. A direct drive motor is connected ____________ to the load it powers.
6. What are some other common power delivery systems?

____

____

____

7. How is pulley size computed?

____

____

____

8. What is the major advantage of a chain and sprocket delivery system over a belt drive?

____

9. How are electric motors controlled?

____

____

____

10. Fuses and circuit breakers are devices that prevent circuit ____

____.

11. Motors are frequently protected by built-in ______________ devices.

12. What are some common types of switches used to control electric motors?

____

____

____

13. List the sensors and other switch actuators described in the textbook. Also list a use for each.

____

____

____

____

____

14. Describe how electric motors are utilized in a system such as a furnace.

____

15. When electric motors are installed or moved, care must be taken to make sure the pulley and belt remain.

____.

Instructor's Score or Approval ____

Job Sheet 36-2

Name ______________________________________ Date ______________

Electric Motor Selection

Objective

Upon completing this activity, you will have demonstrated the ability to select electric motors.

Tools and Materials

Job Sheet 36-2

Pen or pencil

Cooper text

Procedure

Complete the following worksheet.

In Column I, write the number 1 or 2 to indicate whether the motor is suited for an easy starting or hard starting load. In Column II, indicate the starting torque compared with the running torque.

	Column I	*Column II*
Shaded-pole induction	____ 1. Easy start	____ a. ½ to 1 time
Split phase	____ 2. Hard start	____ b. 1 to 1½ times
Permanent-split, capacitor-induction	____	____ c. 2 to 3 times
Soft-start	____	____ d. 3 to 4 times
Capacitor-start, induction-run	____	____ e. ½ to 4½ times
Repulsion-start, induction-run	____	____ f. 4 times
Capacitor-start, capacitor-run	____	____
Repulsion-start, capacitor-run	____	____
Three-phase general purpose	____	____

Instructor's Score or Approval ______________________________________

Job Sheet 36-3

Name ____ Date _______________

Identifying Electrical Control Switches

Objective

Upon completing this activity, you will have demonstrated the ability to identify electrical switch controls.

Tools and Materials

Job Sheet 36-3

Pen or pencil

Textbook

Procedure

Complete the following worksheet by matching the items in Column I with those in Column II.

	Column I	*Column II*
____	1. Float switch	a. Activates a switch after a preset time lapse
____	2. Pilot light	b. Responds to temperature changes
____	3. Timer	c. Glows when the circuit is energized
____	4. Heat sensor	d. Activated by a float
____	5. Flow switch	e. Is activated by the movement of a gas or liquid
____	6. Pressure switch	f. Determines the chemical makeup of a gaseous atmosphere
____	7. Gas analyzer	g. Reacts to changes in pressure

Instructor's Score or Approval ____

Job Sheet 36-4

Name ______________________________ Date ____________

Maintaining Motors, Drives, and Controls

Objective

Upon completing this activity, you will have demonstrated the ability to maintain motors, drives, and controls.

Tools and Materials

Job Sheet 36-4

Motors with drives and controls

Owner's manuals for the motors

Cleaning rags and solvents

Lubricants needed

Procedure

1. Identify the motor by reading the nameplate.
2. Grease shafts or lubricate according to the owner's manual.
3. Clean chains of dirt and grit.
4. Oil chains as recommended.
5. Change the gearbox lubricants as recommended.
6. Adjust belt tensions to the recommended deflection.
7. Check and adjust belt and pulley alignment.
8. Clean the area and put away the materials.

Instructor's Score or Approval ______________________________

SECTION 12

Plumbing, Hydraulic, and Pneumatic Systems

UNIT 37

Plumbing

Water is only one of the materials moved through tubular products. *Plumbing* is the term applied to system pipes and fixtures for water or gas in a building. In agricultural mechanics, the term is applied to all work with pipe and fixtures.

Class Activity 37-1: Understanding Plumbing

Shop Activity 37-2: Connecting Plastic Pipe

Shop Activity 37-3: Repairing Faucets

Shop Activity 37-4: Doing a Plumbing Exercise

Job Sheet 37-1

Name ______________________________ Date ____________

Understanding Plumbing

Objective

Upon completion of this activity, you will have demonstrated an understanding of plumbing.

Tools and Materials

Job Sheet 37-1

Pen or pencil

Procedure

Answer the following questions:

1. What does the term *plumbing* mean in agricultural mechanics?

____

____

2. Define pipe.

____

3. What is a pipe fitting?

____

____

4. List the tools commonly used in plumbing.

____

____

____

5. How is pipe size commonly measured?

____

6. What is galvanized pipe?

____

7. What hazard exists when working with galvanized pipe and heat?

____

____

____

8. Heavy or double-extra-heavy pipe is usually used for construction instead of ____________________.
9. What are the two types of copper pipe?

____

____

10. How are these two types joined?

____

____

11. What do the designations K, L, and M refer to?

12. Which type is required if the pipe is to be buried?

13. What are the most common types of plastic pipe in use today?

14. Why is plastic the most popular type of piping used today?

15. ______ Dripping or hammering may be corrected by tightening or replacing the faucet seal or washer. True or false?

16. What are some common problems with float valves?

17. Why should a septic system be placed far away from a well?

18. Why should septic tanks be pumped out at designated intervals?

Instructor's Score or Approval ______________________

Job Sheet 37-2

Name ____ Date ______________

Connecting Plastic Pipe

Objective

Upon completing this activity, you will have demonstrated the ability to connect plastic pipe.

Tools and Materials

Job Sheet 37-2

PVC pipe, ½"

½" PVC coupling

½" PVC end

PVC cleaner

PVC cement

Miter box

Miter box saw

Knife

Rags

Safety glasses

Procedure

1. Put on the safety glasses.
2. Cut the PVC pipe to 12" in length using the miter box.
3. Remove the rough edges with the knife.
4. Wipe the outside of the pipe ends and the inside of the fittings with a cloth.
5. Clean the outside of the pipe ends and the inside of the fittings with the PVC cleaner.
6. Swab the ends of the pipe and the insides of the fittings with the PVC cement.
7. Insert the pipe into the fitting and give it a quarter turn.
8. Hold the pieces together for 15 seconds while the glue sets.
9. Clean the area and put away the materials.

Instructor's Score or Approval ______________________________________

Job Sheet 37-3

Name ______________________________ Date ____________

Repairing Faucets

Objective

Upon completing this activity, you will have demonstrated the ability to repair a faucet.

Tools and Materials

Job Sheet 37-3

A faucet with a leaking packing nut or drip

Screwdrivers

Adjustable end wrench

O-ring or packing

Faucet washers

Procedure

1. Tighten the packing nut and see if that stops the leak.
2. If the leak does not stop, shut off the water to the faucet.
3. Remove the faucet handle.
4. Remove the packing nut.
5. Replace the O-ring or packing and replace the packing nut.
6. Tighten the packing nut just enough to stop the leak.
7. If the faucet was dripping, remove the faucet stem and replace the faucet washer.
8. Tighten the washer snugly without distortion.
9. Reassemble the faucet.
10. Test for leaks.
11. Clean the area and put away the materials.

Instructor's Score or Approval ______________________________

Job Sheet 37-4

Name ________________________________ Date ______________

Doing a Plumbing Exercise

Objective

Upon completing this activity, you will have demonstrated the ability to do a plumbing exercise using a number of different materials.

Tools and Materials

Job Sheet 37-4

PVC pipe assembly from Job Sheet 37-2

Polyethylene pipe, ½"

½" hose adapter, female (galv.)

½" tee (galv.)

½" coupling (galv.)

½" plug (galv.)

½" sillcock, female (galv.)

½" × 4" nipple (galv.)

½" × 6" nipple (galv.)

Two ½" male adapter inserts

Two ½" stainless steel pipe clamps

½" adapting elbow galvanized to copper

½" copper tee

½" copper union

½" copper adapter, female

½" copper adapter, male

15" type L copper pipe

Flux, pipe compound, solder, solvents, and the tools necessary to complete the exercise

Safety glasses

Procedure

1. Put on the safety glasses.
2. Using the pipe compound, assemble the ½" hose adapter to one end of the galvanized tee.
3. Assemble the 4" nipple to the other end of the tee.
4. Assemble one of the male inserts to the tee.
5. Attach the ½" coupling to the 4" nipple.
6. Assemble the 6" nipple to the coupling.
7. Sweat the male adapter to one end of the copper tee.

8. Cut 6" of copper and sweat it to the other end of the tee.
9. Sweat the female adapter to the tee.
10. Sweat the male part of the union to the 6" pipe.
11. Sweat the female part of the union to a copper pipe cut to equal the measurement of the galvanized assembly.
12. Slip the compression ring onto the pipe and sweat the adapting elbow onto the copper pipe.
13. Assemble the other male insert to the adapting elbow.
14. Connect the male insert and 12" of polyethylene pipe using the stainless steel clamp.
15. Assemble the other end of the 12" polyethylene pipe to the male adapter (already on the galvanized tee) with the other stainless steel clamp.
16. Connect the sillcock to the male adapter in the copper tee.
17. Assemble the ½" PVC pipe assembly (from Job Sheet 37-2) to the 6" galvanized nipple.
18. Align and assemble the copper union.
19. Assemble the ½" galvanized plug into the ½" copper female adapter.
20. Attach the hose and check for leaks.
21. Clean the area and put away the materials.

Instructor's Score or Approval ______________________________

UNIT 38

Irrigation Technology

Irrigation has been used by many civilizations over the centuries. This practice has enabled higher production and reclamation of otherwise less-productive land.

Class Activity 38-1: Identifying Irrigation Systems

Class Activity 38-2: Understanding Irrigation

Job Sheet 38-1

Name ______________________________________ Date ______________

Identifying Irrigation Systems

Objective

Upon completing this activity, you will have demonstrated the ability to identify different irrigation systems.

Tools and Materials

Job Sheet 38-1

Pen or pencil

Procedure

Match each irrigation system with its description.

____	1. Surface	a. Simulated rain
____	2. Subirrigation	b. By gravity
____	3. Trickle	c. By using subsoil
____	4. Sprinkler	d. By using small emitters

Instructor's Score or Approval ______________________________________

Job Sheet 38-2

Name ______________________________ Date ____________

Understanding Irrigation

Objective

Upon completing this activity, you will have demonstrated an understanding of irrigation.

Tools and Materials

Job Sheet 38-2

Pen or pencil

Procedure

Answer the following questions:

1. Define irrigation.

____

____

2. What are the benefits of irrigation?

____

____

____

____

____

3. List the 12 questions to consider before installing an irrigation system.

____

____

____

____

____

____

____

____

____

____

____

____

4. To apply 1" of water over an acre of soil requires ____________ gallons of water.
5. Define seasonal demand.

____

____

____

6. How do daily water requirements for alfalfa grown in Virginia compare with those grown in southern Texas?

____

____

7. Explain the term *peak-use*.

____

____

____

8. ______ Before starting an irrigation system, you should check local water-use laws. True or false?
9. ______ Irrigation wells and pumps must be designed to fit a given installation. True or false?
10. ______ Planning irrigation systems requires no knowledge of plant growth. True or false?
11. What are moisture sensors?

____

____

____

Instructor's Score or Approval ______________________________

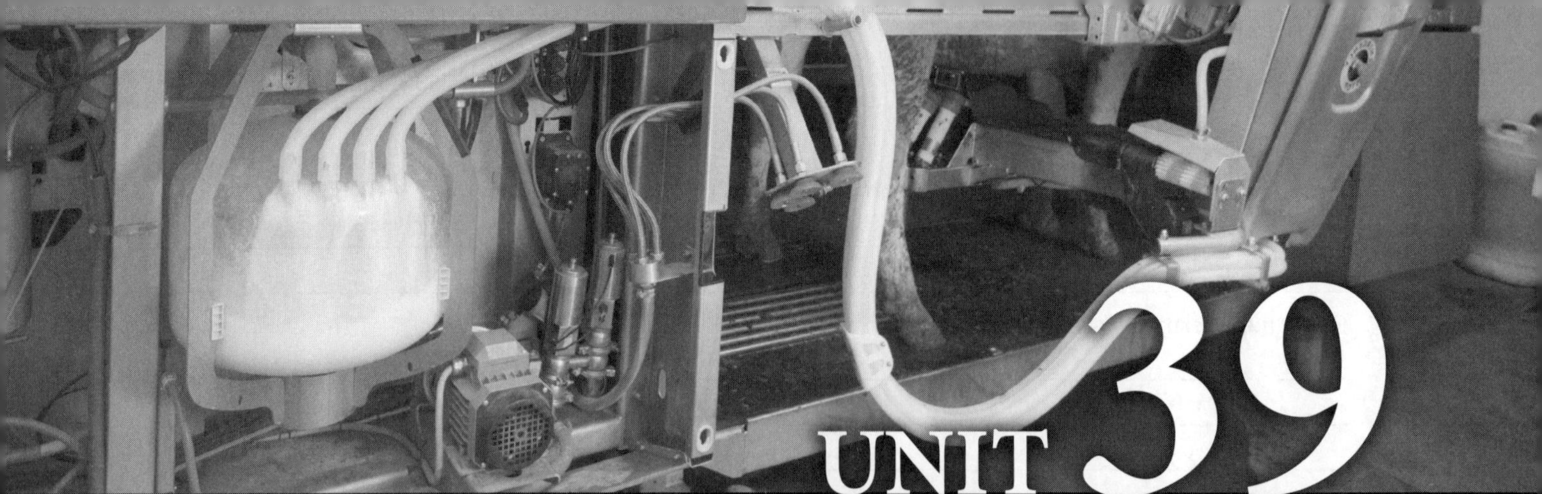

UNIT 39

Hydraulic, Pneumatic, and Robotic Power

Fluid power is widely used in many industries, including agriculture. Fluids include liquids and gases. Fluid power technology is discussed in the following activities.

Class Activity 39-1: Understanding Pneumatic, Hydraulic, and Robotic Power

Class Activity 39-2: Identifying Fluid Additives

Class Activity 39-3: Identifying Hydraulic System Maintenance Items

Class Activity 39-4: Identifying Robotic Work Areas

Job Sheet 39-1

Name ______________________________ Date ____________

Understanding Pneumatic, Hydraulic, and Robotic Power

Objective

Upon completing this activity, you will have demonstrated an understanding of pneumatics, hydraulics, and robotic power.

Tools and Materials

Job Sheet 39-1

Pen or pencil

Procedure

Answer the following questions:

1. Define hydraulics.
____
____
2. Define pneumatics.
____
____
3. Hydraulics and pneumatics together are known as ______________ power.
4. ______ A log splitter uses pneumatic power. True or false?
5. ______ A log splitter would be an example of fluid power. True or false?
6. Define force.
____
____
7. Define pressure.
____
____
8. ______ Pressure in liquid containers is exerted equally on all surfaces. True or false?
9. ______ When equal pressure is applied to cylinder A, the resulting force will be greater in cylinder B if cylinder B is larger in diameter. True or false?
10. ______ Boyle's Law describes why basketballs bounce. True or false?
11. ______ As the diameter of a pipe decreases, the pressure increases. True or false?
12. What is the meaning of the SAE rating of 10W40?
____
____
____
13. What is the meaning of an API rating?
____
____
____

14. List the components of a hydraulic system.

15. Name the three types of hydraulic pumps.

16. What is a fluid coupling?

17. List the characteristics of a quality hydraulic hose.

18. Describe the function of a pressure regulator.

19. What is a check (or direction-control) valve?

20. Describe an equipment control valve.

21. Why are filtering systems needed within a hydraulic system?

22. What is a full-flow filtering system?

23. What is a bypass filtering system?

24. Describe a hydraulic cylinder.

____

____

____

25. Describe a hydraulic motor.

____

____

____

26. Describe the maintenance needed by the following hydraulic components:

Fluid level

____

Fluid filters

____

Quick couplings

____

27. What are the signs that a hydraulic fluid change is needed?

____

____

____

28. _______ Pneumatic system components are often similar to hydraulic system components. True or false?

29. What is the biggest difference between hydraulic cylinders and pneumatic cylinders?

____

30. Define robotics.

____

____

31. Robots are different from other machines because they ____

____

____

____.

32. Robots are well suited to what kinds of tasks?

____

____

____

33. Robotic motions include ____________________ and ____________________.

34. Why is electric power commonly used to power robots?

____

____

Instructor's Score or Approval ____

Job Sheet 39-2

Name ____ Date _______________

Identifying Fluid Additives

Objective

Upon completing this activity, you will have demonstrated the ability to identify fluid additives.

Tools and Materials

Job Sheet 39-2

Pen or pencil

Procedure

Match each fluid additive with its description.

____	1. Viscosity improver	a. Prevents oil breakdown
____	2. Antifoam	b. Helps polish moving parts
____	3. Corrosion inhibitor	c. Reduces or prevents rust
____	4. Rust inhibitor	d. Provides stability with temperature change
____	5. Antiscuff	e. Reduces foaming
____	6. Extreme-pressure resistor	f. Reduces or prevents corrosion

Instructor's Score or Approval ____

Job Sheet 39-3

Name ______________________ Date ____________

Identifying Hydraulic System Maintenance Items

Objective

Upon completing this activity, you will have demonstrated the ability to identify hydraulic system maintenance items.

Tools and Materials

Job Sheet 39-3

Pen or pencil

Procedure

Match the hydraulic component with the correct maintenance procedure.

____	1. Fluid level	a. Drain and refill at the manufacturer's time interval
____	2. Fluid filters	b. Keep clean by capping when not in use
____	3. Quick-couplings	c. Checked frequently and maintained
____	4. Fluid changes	d. Unscrew and replace at specified intervals

Instructor's Score or Approval ______________________

Job Sheet 39-4

Name ______________________________ Date ____________

Identifying Robotic Work Areas

Objective

Upon completing this activity, you will have demonstrated the ability to identify the robotic work areas.

Tools and Materials

Job Sheet 39-4

Pen or pencil

Procedure

Match each work area with the correct description.

____ 1. Cartesian	a. Three rotational axes
____ 2. Cylindrical	b. Two rotational and one translational axes
____ 3. Hollow sphere	c. Box-like work area
____ 4. Solid sphere	d. One rotational and two translational axes

Instructor's Score or Approval ______________________________

SECTION 13

Concrete and Masonry

Unit 40

Concrete and Masonry

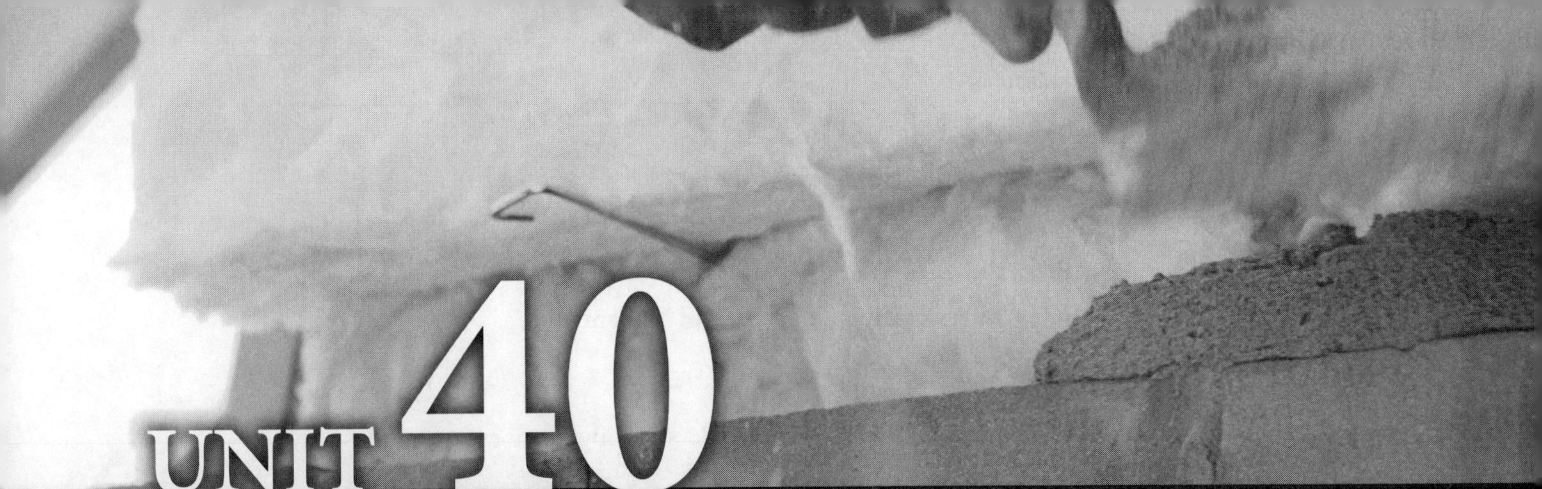

UNIT 40

Concrete and Masonry

Concrete and masonry materials are very widely used. Knowledge of concrete construction is an essential skill in agricultural mechanics because of its many uses and the flexibility of its forming and placement.

Class Activity 40-1: Understanding Concrete and Masonry

Shop Activity 40-2: Testing Sand for Silt or Clay

Class Activity 40-3: Estimating Materials

Shop Activity 40-4: Preparing Forms

Shop Activity 40-5: Laying a Concrete Block Corner

Job Sheet 40-1

Name ______________________________ Date ____________

Understanding Concrete and Masonry

Objective

Upon completing this activity, you will have demonstrated an understanding of concrete and masonry.

Tools and Materials

Job Sheet 40-1

Pen or pencil

Procedure

Answer the following questions:

1. Define masonry.

2. Define cement.

3. Define concrete.

4. What is mortar?

5. What are the advantages of using concrete as a building material?

6. Why does quality concrete require the use of graded aggregates?

7. Why should sand be tested for silt or clay?

8. What are the characteristics of a workable concrete mix?

9. What are concrete forms?

10. What materials are usually used in form construction?

11. What is a construction joint? A control joint?

12. Why and how is concrete reinforced?

13. Identify the concrete finishing tools pictured in the following diagram:

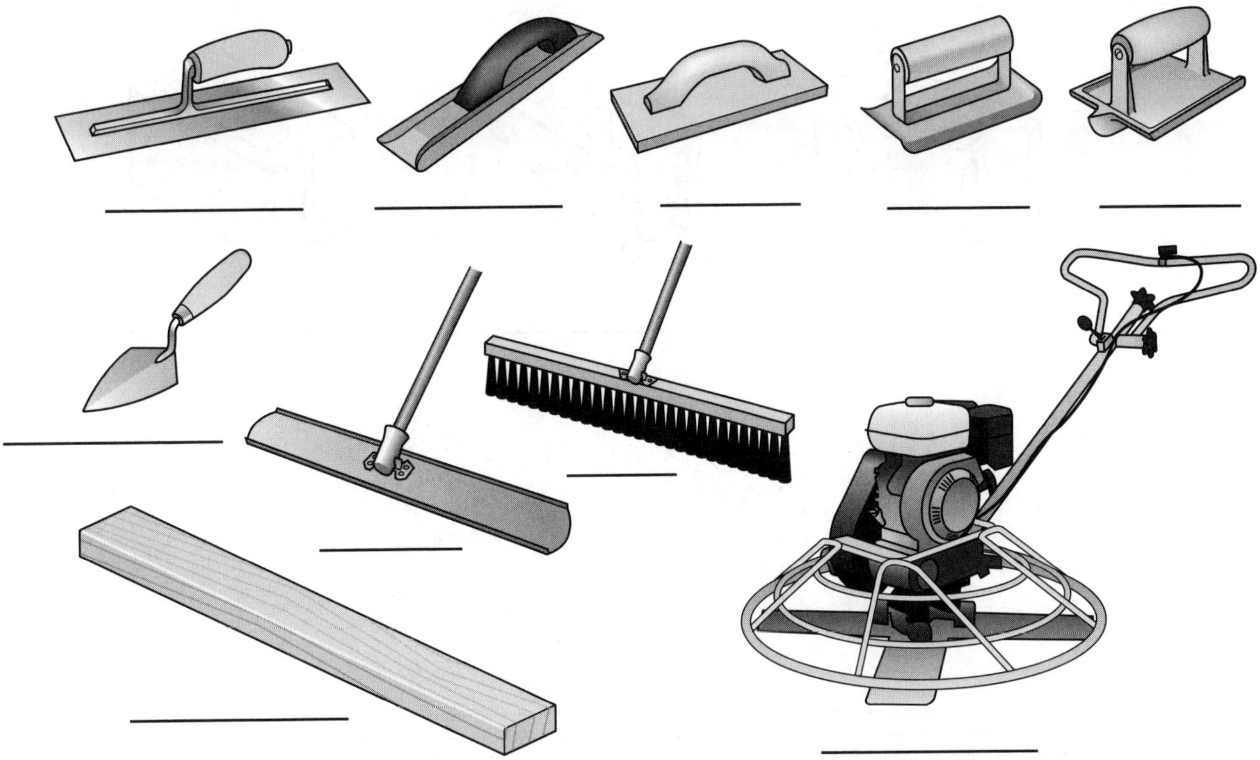

Figure 40-1 Concrete Tools

14. How soon does concrete start to harden after mixing is stopped?

15. What is screeding? Why is it needed?

16. What is floating? Why is it done?

17. What is air-entrained concrete? Why is it used?

18. What is a broom finish? Why is it used?

19. What is curing? Why is moisture important?

20. Identify the common types of blocks shown in the following diagram:

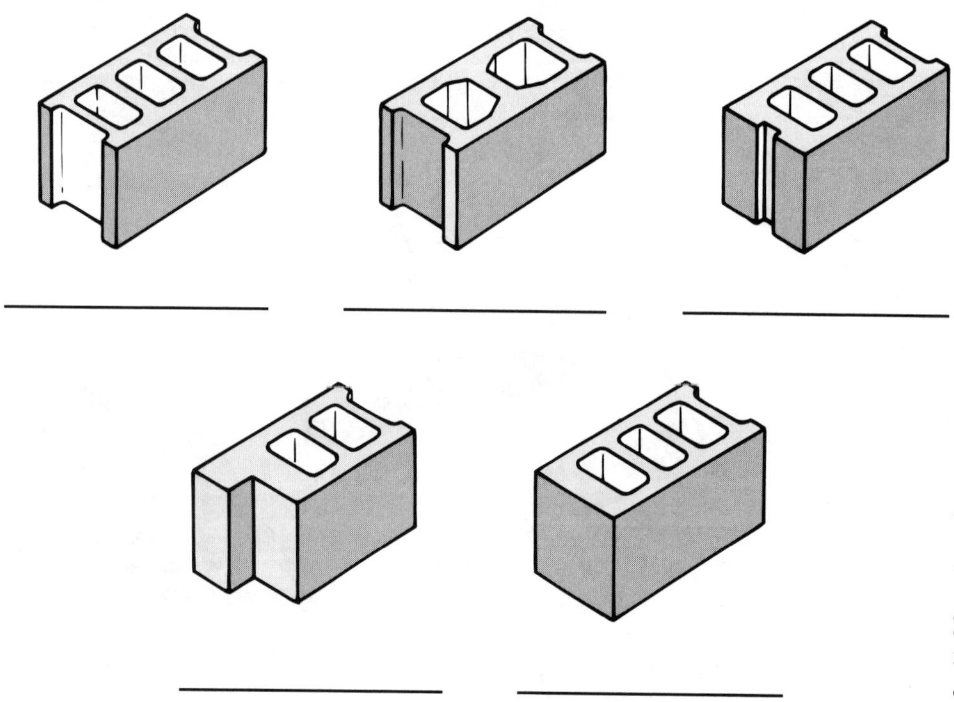

Figure 40-2 Types of Concrete Blocks

21. What is the nominal size of a concrete block?
22. How many blocks will be needed for a wall 36' long and 8' high with a 2'8" × 6'8" door and a 32" × 48" window?

____

____

23. What is a footer? How deep should it be placed?

____

____

24. Why must the wetness of mortar be carefully controlled?

____

____

25. How are blocks installed straight, level, and plumb?

____

____

____

____

Instructor's Score or Approval ______________________________

Job Sheet 40-2

Name ____ Date ______________

Testing Sand for Silt or Clay

Objective

Upon completing this activity, you will have demonstrated the ability to test sand for silt or clay.

Tools and Materials

Job Sheet 40-2

Quart fruit jar

Sample of sand to be tested

Water

Ruler

Procedure

1. Place 2" of sand in the quart jar.
2. Add water so that the jar is three-fourths full.
3. Cap the jar and shake vigorously for 1 minute.
4. Shake the jar sideways to level the sand.
5. Set the jar aside for 1 hour.
6. Measure the silt at the top of the sand.
7. If the silt layer is more than ⅛", reject the sand or wash it.
8. If the layer is less than ⅛", let it set overnight to check the clay content.
9. If the clay layer is more than ⅛", reject the sand or wash it.
10. A clay layer of less than ⅛" means that the sand may be used.
11. Clean the area and put away the materials.

Instructor's Score or Approval ____

Job Sheet 40-3

Name ______________________ Date ____________

Estimating Materials

Objective

Upon completing this activity, you will have demonstrated the ability to estimate materials needed for a concrete job.

Tools and Materials

Job Sheet 40-3

Pen or pencil

Procedure

Determine the materials for a sidewalk 4' wide × 40' long × 4" thick. Use a 1, 2½, 3½ mix (see text Figure 40-9).

1. Compute the cubic feet in the sidewalk. ____________ cubic feet
2. Compute the cubic yards in the sidewalk. ____________ cubic yards
3. Compute the amount of cement needed. ____________ pounds or cubic feet
4. Compute the amount of fine aggregate needed. ____________ pounds or cubic feet
5. Compute the amount of coarse aggregate needed. ____________ pounds or cubic feet

Optional: Compute the cost of the materials.

Instructor's Score or Approval ______________________

Job Sheet 40-4

Name ______________________________ Date ____________

Preparing Forms

Objective

Upon completing this activity, you will have demonstrated the ability to construct concrete forms.

Tools and Materials

Job Sheet 40-4

2" × 4" material for forms and stakes

Sledgehammer

Claw hammer

8d nails

Level

Table saw

Carpenter's square

Shovel

Handsaw

Reasonably level area for construction

Safety glasses

Procedure

1. Put on the safety glasses.
2. Locate one corner of the form your instructor has chosen.
3. Cut the 2" × 4" to make the form ends and sides needed.
4. Cut and sharpen a stake for each corner and one for every 3' of form length and width.
5. Set one side of the form and stake it every 3'.
6. Level the side and nail through the form into the stakes.
7. Establish a square corner with the carpenter's square and then stake the end.
8. Level the end to the side and then nail into the stakes.
9. Build the remaining side and end in the same manner.
10. Saw off any stakes that extend above the top of the forms.
11. Clean the area and put away the materials.

Note: If concrete were to be poured in these forms, the surface within the forms would need to be leveled to the depth of the forms.

Instructor's Score or Approval ______________________________

Job Sheet 40-5

Name ______________________________ Date ____________

Laying a Concrete Block Corner

Objective

Upon completing this activity, you will have demonstrated the ability to lay a concrete block corner.

Tools and Materials

Job Sheet 40-5

Mixing hoe

Mortar box

Chalk line

Mortar mix

Level with a plumb

Water

21 stretcher blocks

Carpenter's square

7 corner blocks

Story pole or stakes

Mason's trowel

Procedure

1. Establish the corner with the chalk line and square.
2. Mix the mortar to a stiff consistency.
3. Spread a layer of mortar on the footer.
4. Set a corner block precisely on the corner and level it to the chalk line.
5. Set 3 stretcher blocks on end and apply mortar to the ears with a wiping motion.
6. Set the mortared ends against the first block and use the handle of the trowel to level and plumb them to the left of the first block.
7. Spread a mortar bed to the right of the corner block and repeat Step 6.
8. Move the chalk lines up 8".
9. Place a mortar bed across the first corner block and the stretcher block beside it.
10. Set the next corner block across these two blocks and level and plumb it to the chalk line.
11. Complete the mortar bed and set stretcher blocks as in Steps 5 and 6.
12. Repeat on the other side of the corner.
13. Repeat Steps 8 to 12 until the corner is complete, remembering that each course uses one fewer block.

14. Cut the excess mortar off with the trowel.
15. After the mortar dries a bit, finish the joint with a block or jointer.
16. Remove and clean the blocks before the mortar hardens completely.
17. Clean the area and put away the materials.

Instructor's Score or Approval ______________________________

SECTION 14

Agricultural Structures

Unit 41

Planning and Constructing Agricultural Structures

Unit 42

Aquaculture, Greenhouse, and Hydroponics Structures

Unit 43

Fence Design and Construction

UNIT 41
Planning and Constructing Agricultural Structures

Many skills are utilized in constructing and maintaining agricultural buildings or structures. These skills are also used in other industries.

Class Activity 41-1: Understanding Agricultural Structures

Class Activity 41-2: Identifying Veneer Grades

Shop Activity 41-3: Laying Out a Building

Shop Activity 41-4: Laying Out Stairs

Shop Activity 41-5: Laying Out a Rafter

Shop Activity 41-6: Tying a Square Knot

Shop Activity 41-7: Tying a Bowline Knot

Shop Activity 41-8: Tying a Clove Hitch

Shop Activity 41-9: Tying Two Half Hitches

Shop Activity 41-10: Tying a Timber Hitch

Shop Activity 41-11: Tying a Sheet Bend

Job Sheet 41-1

Name ______________________________ Date ____________

Understanding Agricultural Structures

Objective

Upon completing this activity, you will have demonstrated an understanding of agricultural structures.

Tools and Materials

Job Sheet 41-1

Pen or pencil

Procedure

Answer the following questions:

1. Planning is important because it saves both ____________ and ____________.
2. List the principles of farmstead layout.

   ____

   ____

3. Name a farm building for which professional help in design is recommended.

   ____

4. What types of structures might a farm family be able to plan on their own?

   ____

   ____

5. Define rafter.

   ____

   ____

6. Define girder.

   ____

   ____

7. Define truss.

   ____

   ____

8. Buildings are classified by their ____________ or the type of ______________________ used.
9. A pole building is supported by ____________.
10. The most common materials used for agricultural buildings are ______________________.

    ____

11. Roofing and siding for agricultural buildings are usually made of ____________ or ____________.
12. What is OSB?

    ____

    ____

13. Why is OSB important?

____

14. The most commonly used insulation materials include ____
____.

15. When livestock buildings are insulated, _______________ is very important.

Instructor's Score or Approval ____

Job Sheet 41-2

Name ______________________________________ Date ______________

Identifying Veneer Grades

Objective

Upon completing this activity, you will have demonstrated the ability to identify veneer lumber grades.

Tools and Materials

Job Sheet 41-2

Pen or pencil

Procedure

Match the items in Column I with those in Column II.

Column I	*Column II*
____ 1. A	a. Improved C veneer; splits limited to ⅛" and knots and borer holes limited to ¼" × ½"
____ 2. B	b. Tight knots to 1½" or knotholes to 1" across grain
____ 3. C	c. Knots and knotholes to 2½" across grain
____ 4. C plugged	d. Smooth and paintable with limited repairs permitted with the grain
____ 5. D	e. Solid surface with shims and circular repair plugs and tight knots to 1"

Instructor's Score or Approval ______________________________________

Job Sheet 41-3

Name ____ Date ______________

Laying Out a Building

Objective

Upon completing this activity, you will have demonstrated the ability to lay out a building.

Tools and Materials

Job Sheet 41-3

Line level or transit

16 – 2" × 2" × 16" stakes

8 – 1" × 4" × 3' boards

Hammer

Nails

Chalk lines

Sledgehammer

Safety glasses

Procedure

1. Put on the safety glasses.
2. Locate one corner of the proposed building and set one stake, driving a nail into the top of the stake as a reference point.
3. Measure the length of the building in the desired direction from this stake and set the second stake level with the first, using the line level or transit.
4. Drive a nail into the top of the stake. This nail should be exactly the length of the building from the nail in the first stake.
5. Measure the width of the building perpendicular to the length of the building established in Steps 3 and 4 (use the 3-4-5 right-triangle method).
6. Set the third temporary stake and level it with the first stake.
7. Drive a nail in the stake at the exact length of the end of the building.
8. Measure the length of the building parallel to the first side laid out.
9. Measure the second end of the building from the second stake and set a stake where the line crosses the line made in Step 8. Level the fourth stake with the other three.
10. Drive a nail into the fourth stake at the exact location where the distance from stake 2 and the distance from stake 3 intersect.
11. Measure the distance from the fourth nail to the first and from the second nail to the third.
12. If the distances are unequal, shift the third and fourth stakes toward or away from each other as needed.
13. With the stakes leveled and the nails at the proper distances, install batter boards a foot beyond each stake.
14. Level the tops of the batter boards with the stakes.

15. Use string to run from the batter boards directly over the tops of the nails placed in the stakes. Repeat this process four times until the four walls of the building are represented with string.
16. Use saw kerfs or nails to identify the exact location of the string on each batter board.
17. Remove the stakes and the string and use the kerfs or nails on the batter boards to reference the corners.
18. Clean the area and put away the materials.

Instructor's Score or Approval ______________________________

Job Sheet 41-4

Name ______________________________ Date ____________

Laying Out Stairs

Objective

Upon completing this activity, you will have demonstrated the ability to lay out stairs.

Tools and Materials

Job Sheet 41-4

2" × 10" × 10' board

Carpenter's square

Pencil

Two sawhorses

2' bench rule

Two small C-clamps

Procedure

1. Lay the 2" × 10" board on the sawhorses.
2. Compute the risers and treads, assuming a 60" rise.
3. Clamp the bench rule on the carpenter's square with the tread measure on the blade and the riser measure on the tongue.
4. Place the carpenter's square on the 2" × 10" board with the tongue at the left end.
5. Adjust the square along the edge until the full riser minus the thickness of the tread can be marked at the end of the board.
6. Reverse the square and place the blade at this mark. Then, mark the tread and riser cuts.
7. Slide the square to the riser mark and mark the next cuts.
8. Repeat Step 7 until all cuts are marked.
9. Clean the area and put away the materials.

Instructor's Score or Approval ______________________________

Job Sheet 41-5

Name ______________________________ Date ____________

Laying Out a Rafter

Objective

Upon completing this activity, you will have demonstrated the ability to lay out a rafter.

Tools and Materials

Job Sheet 41-5

Carpenter's square

2" × 6" × 10' board

Pencil

Two sawhorses

2' bench rule

Procedure

Assume a run of 6', a 6⁄12 slope, and an overhang of 6". (Figure 41-28 in the textbook explains this procedure.)

1. Lay the 2" × 6" board on the sawhorses.
2. Lay out a marking line 2" from the bottom edge of the rafter that runs the length of the rafter.
3. Lay the square on the rafter so the tongue is near the left end. Put the 6" mark of the tongue and the 12" mark of the blade both on the marking line.
4. Adjust the square until it is within ¼" of the end of the rafter, and mark the plumb cut.
5. While keeping the square on the marking line, put a dot on the rafter where the 16" mark of the blade intersects the bottom edge of the rafter. This mark is for the point of the bird's mouth.
6. Align the square so that the left side of the tongue is on the dot made in Step 5 and the 6" and 12" marks are still on the marking line. Mark a plumb line from the marking line to the bottom of the rafter for the bird's mouth.
7. Turn the square 180° and square the bird's mouth by putting the 12" mark of the blade on the plumb line made in Step 6 and the 6" mark of the tongue on the marking line.
8. Calculate the length of the rafter and measure the distance along the marking line from the point of the bird's mouth. Place a dot on the marking line at this location.
9. Mark a plumb cut parallel to the first plumb cut marked in Step 4.
10. Clean the area and put away the materials.

Note: You could also step off the rafter by using the technique for laying out stairs.

Instructor's Score or Approval ______________________________

Job Sheet 41-6

Name ______________________________ Date ____________

Tying a Square Knot

Objective

Upon completing this activity, you will have demonstrated the ability to tie a square knot.

Tools and Materials

Job Sheet 41-6

2' of rope

Procedure

1. Hold the ends of the rope in your left hand with the ends toward the outside.
2. Bring the right end under and the left end under and loop the ends back to the center.
3. With the end on the left in front, wrap the rope around and pull it through the loop formed.
4. With the other end, bring the rope over the first and pull it through the loop formed.
5. Check that the result looks like Figure 41-31 in the textbook.
6. Practice the knot until it is easy to complete.
7. Move on to the next exercise.

Instructor's Score or Approval ______________________________

Job Sheet 41-7

Name ______________________________ Date ____________

Tying a Bowline Knot

Objective

Upon completing this activity, you will have demonstrated the ability to tie a bowline knot.

Tools and Materials

Job Sheet 41-7

Rope used in Job Sheet 41-6

Procedure

1. Hold the rope in the middle with your left hand.
2. Form a 2" to 3" circle with your right hand and bring the rope back over the top of the end in your left hand.
3. Form a larger circle and bring the right end up through the first circle.
4. Move this end under the rope in your left hand and around the rope.
5. Bring the end down through the first circle and tighten.
6. Check to see that it looks like Figure 41-31 in the textbook.
7. Practice the knot several times.
8. Continue to the next exercise.

Instructor's Score or Approval ______________________________

Job Sheet 41-8

Name ____ Date _______________

Tying a Clove Hitch

Objective

Upon completing this activity, you will have demonstrated the ability to tie a clove hitch.

Tools and Materials

Job Sheet 41-8

Rope used in the previous activity

Vise

Short handle

Procedure

1. Place the short handle in the vise so it extends a foot or more vertically.
2. Using the rope, form a 3" circle with the left end on top.
3. Place this circle over the short handle.
4. Form a second circle with the right end underneath.
5. Drop the circle over the handle and tighten (see text Figure 41-31).
6. Practice the knot several times.
7. Proceed to the next exercise.

Instructor's Score or Approval ____

Job Sheet 41-9

Name ______________________________ Date ____________

Tying Two Half Hitches

Objective

Upon completing this activity, you will have demonstrated the ability to tie two half hitches.

Tools and Materials

Job Sheet 41-9

The rope used for the previous exercises

Vise

Short handle used for the previous exercise

Procedure

1. Set the short handle in the vise.
2. With one end of the rope in your left hand, pass the right end around the handle from the back.
3. Pass the rope over the end in your left hand and bring the end around and up through the loop around the handle.
4. Loop the rope around the part in your left hand again and pull it up through the resulting loop.
5. Tighten the ends.
6. Compare it to Figure 41-31 in the textbook.
7. Practice the knot several times.
8. Continue to the next exercise.

Instructor's Score or Approval ______________________________

Job Sheet 41-10

Name ______________________________ Date ____________

Tying a Timber Hitch

Objective

Upon completing this activity, you will have demonstrated the ability to tie a timber hitch.

Tools and Materials

Job Sheet 41-10

The rope used in the previous exercise

Vise

Short handle

Procedure

1. Set the handle in the vise.
2. Wrap the rope around the handle loosely in a clockwise direction below the other end.
3. Hold this rope up along the handle and bring the other end around and under itself twice and tighten.
4. With the upper end, form a second loop around the handle in a counterclockwise direction and pull it underneath the start of the loop.
5. Compare the result to Figure 41-31 in the textbook.
6. Practice the knot several times.
7. Proceed to the next exercise.

Instructor's Score or Approval ______________________________

Job Sheet 41-11

Name ______________________________ Date ____________

Tying a Sheet Bend

Objective

Upon completing this activity, you will have demonstrated the ability to tie two different-sized ropes together.

Tools and Materials

Job Sheet 41-11

2' of a smaller-sized rope

Rope used in the previous exercises

Procedure

1. Form a loop in the larger rope and hold the two ends in your left hand.
2. Pass one end of the smaller rope through the loop and around both sides of the loop.
3. Pass the end of the smaller rope under itself as it comes through the loop.
4. Compare this knot to Figure 41-31 in the textbook.
5. Practice the knot several times.
6. Clean the area and put away the materials.

Instructor's Score or Approval ______________________________

UNIT 42 Aquaculture, Greenhouse, and Hydroponics Structures

Aquaculture and hydroponics production have joined greenhouse production as growing areas of agricultural enterprises.

Class Activity 42-1: Understanding Aquaculture, Greenhouse, and Hydroponics Structures

Class Activity 42-2: Identifying Aquaculture Systems

Class Activity 42-3: Identifying Greenhouse Types

Class Activity 42-4: Identifying Hydroponics Systems

Job Sheet 42-1

Name ______________________ Date ____________

Understanding Aquaculture, Greenhouse, and Hydroponics Structures

Objective

Upon completing this activity, you will have demonstrated an understanding of aquaculture, greenhouse, and hydroponics structures.

Tools and Materials

Job Sheet 42-1

Pen or pencil

Procedure

Answer the following questions:

1. Describe an open aquaculture system.

____

____

____

2. Describe a semi-open system.

____

____

____

3. What additional actions must the grower take in using a semi-open system?

____

____

____

____

____

4. What are the two ways oxygen is added to an aquaculture system?

____

____

5. Where are closed aquaculture systems used?

____

____

6. What are the major components of a closed system?

____

____

____

7. What is the major advantage of a round tank?
____
____

8. What is the advantage of a rectangular tank?
____
____

9. What is the advantage of a plastic liner?
____
____

10. What is the disadvantage of a plastic liner?
____
____

11. Auxiliary equipment capacity is based on fish __________ in the system.

12. Describe a greenhouse.
____
____
____

13. What are commonly used greenhouse covering materials?
____
____
____

14. In summer, the greenhouse must be __________ or __________ to control the temperature.

15. Greenhouse location should provide full __________ to the sun.

16. What factors determine wintertime heating requirements in a greenhouse?
____
____
____

17. Define hydroponics.
____
____
____

18. What are the components of a hydroponics growing system?
____
____
____

19. ______ The hydroponics system produces the highest production of any plant culture system. True or false?

20. ______ In the hydroponics system, very little supervision is required. True or false?

Instructor's Score or Approval ______________________________

Job Sheet 42-2

Name ______________________________ Date ____________

Identifying Aquaculture Systems

Objective

Upon completing this activity, you will have demonstrated the ability to identify aquaculture systems.

Tools and Materials

Job Sheet 42-2

Pen or pencil

Procedure

Match the items in Column I with those in Column II.

	Column I	*Column II*
____	1. Open system	a. Recirculating system
____	2. Semi-open system	b. Low density
____	3. Closed system	c. Pumps and flushing

Instructor's Score or Approval ______________________________

Job Sheet 42-3

Name ________________________________ Date ____________

Identifying Greenhouse Types

Objective

Upon completing this activity, you will have demonstrated the ability to identify greenhouse types.

Tools and Materials

Job Sheet 42-3

Pen or pencil

Procedure

Match the greenhouse structures listed in Column I with their descriptions listed in Column II.

	Column I	*Column II*
____	1. Gable	a. Rounded top
____	2. Sawtooth	b. Single-peaked
____	3. Quonset	c. Irregular top line
____	4. Lean-to	d. Added on to another building

Instructor's Score or Approval ________________________________

Job Sheet 42-4

Name ________________________________ Date ____________

Identifying Hydroponics Systems

Objective

Upon completing this activity, you will have demonstrated the ability to identify hydroponics systems.

Tools and Materials

Job Sheet 42-4

Pen or pencil

Procedure

Match the items in Column I with those in Column II.

	Column I	*Column II*
____	1. Bag culture	a. In continuously circulating nutrient solutions
____	2. Trough culture	b. In bags with emitters
____	3. Tube culture	c. In intermittent nutrient circulation

Instructor's Score or Approval ________________________________

UNIT 43

Fence Design and Construction

Fences, which have been used for thousands of years, come in a wide variety of types and have many uses. A properly constructed fence can last for many years.

Class Activity 43-1: Understanding Early Uses of Fences

Class Activity 43-2: Understanding Modern Fences

Class Activity 43-3: Identifying Fences

Class Activity 43-4: Understanding Steps in Constructing a Fence

Job Sheet 43-1

Name ______________________________ Date ____________

Understanding Early Uses of Fences

Objective

Upon completing this activity, you will have demonstrated an understanding of the early uses of fences.

Tools and Materials

Job Sheet 43-1

Pen or pencil

Procedure

Answer the following questions:

1. When people first began to grow crops, why did they start building fences?

____

____

____

2. What were these fences made of?

____

____

____

3. After animals became domesticated, what were fences used for?

____

____

____

4. What was used early on as a fence-building material?

____

____

____

5. Where can you still find these fences today?

____

____

____

6. What other materials were used to build fences?

____

____

____

7. What is barbwire?

8. What were some of the disadvantages associated with barbwire?

9. What were the "range wars"?

Instructor's Score or Approval

Job Sheet 43-2

Name ________________________________ Date ____________

Understanding Modern Fences

Objective

Upon completing this activity, you will have demonstrated the ability to understand modern fences.

Tools and Materials

Job Sheet 43-2

Pen or pencil

Procedure

Answer the following questions:

1. What are the five reasons to build a fence today?

2. What are landscape fences used for?

3. What are the two types of landscape fence?

4. What are the characteristics of a wooden fence?

5. What is the newest type of fencing material in use today?

6. Name five types of fencing in use today:

7. What are the advantages and disadvantages of an electric fence?

8. What is a lightning arrester used for?

9. What is the most common type of fence used today?

10. What are the differences among the various types of wire available for fencing?

11. What are the differences between barbwire and high-tensile wire?

12. What is a major disadvantage of high-tensile wire, and how can you combat it?

13. How is woven wire classified?

Instructor's Score or Approval ______________________

Job Sheet 43-3

Name ______________________________ Date ____________

Identifying Fences

Objective

Upon completing this activity, you will have demonstrated the ability to identify the different kinds of fences.

Tools and Materials

Job Sheet 43-3

Pen or pencil

Procedure

Name each type of fence used today and describe its characteristics.

1.

2.

3.

4.

5.

6.

Instructor's Score or Approval ______________________________

Job Sheet 43-4

Name ______________________________ Date ____________

Understanding Steps in Constructing a Fence

Objective

Upon completing this activity, you will have demonstrated an understanding of how to construct a fence.

Tools and Materials

Job Sheet 43-4

Pen or pencil

Procedure

Answer the following questions:

1. What are the safety guidelines for constructing a fence?

____

____

____

2. List the five steps used in constructing a fence and describe the procedure for each.

____

____

____

____

____

Instructor's Score or Approval ______________________________